THE OCEAN BASINS:
THEIR STRUCTURE AND EVOLUTION

THE OCEANOGRAPHY COURSE TEAM

Authors
Joan Brown
Angela Colling
Dave Park
John Phillips
Dave Rothery
John Wright

Editor
Gerry Bearman

Design and Illustration
Sue Dobson
Ray Munns
Ros Porter
Jane Sheppard

This Volume forms part of an Open University course. For general availability of all the Volumes in the Oceanography Series, please contact your regular supplier, or in case of difficulty the appropriate Pergamon office.

Further information on Open University courses may be obtained from: The Admissions Office, The Open University, P.O. Box 48, Walton Hall, Milton Keynes MK7 6AA.

Cover illustration: Satellite photograph showing distribution of phytoplankton pigments in the North Atlantic off the US coast in the region of the Gulf Stream and the Labrador Current. *(NASA and O. Brown and R. Evans, University of Miami.)*

THE OCEAN BASINS: THEIR STRUCTURE AND EVOLUTION

PREPARED BY AN OPEN UNIVERSITY COURSE TEAM

PERGAMON PRESS
OXFORD · NEW YORK · SEOUL · TOKYO

in association with

THE OPEN UNIVERSITY
WALTON HALL, MILTON KEYNES MK7 6AA, ENGLAND

U.K.	Pergamon Press Ltd, Headington Hill Hall, Oxford OX3 0BW, England
U.S.A.	Pergamon Press, Inc., 660 White Plains Road, Tarrytown, New York 10591-5153, U.S.A.
KOREA	Pergamon Press Korea, KPO Box 315, Seoul 110-603, Korea
JAPAN	Pergamon Press Japan, Tsunashima Building Annex, 3-20-12 Yushima, Bunkyo-ku, Tokyo 113, Japan

First edition 1989
Reprinted with corrections 1991, 1992

Library of Congress Cataloging in Publication Data

The Ocean Basins.
1. Submarine geology. 2. Palaeoceanography.
I. Open University.
QE39.023   1988   551.46′08   88–17879

British Library Cataloguing in Publication Data

The Ocean Basins.
1. Oceans. Bed. Mechanics
I. Open University. *Oceanography Course Team*
551.46′08
ISBN 0–08–036366–0 Hardcover
ISBN 0–08–036365–2 Flexicover

Jointly published by the Open University, Walton Hall, Milton Keynes MK7 6AA and Pergamon Press Ltd, Headington Hill Hall, Oxford OX3 0BW.

Designed by the Graphic Design Group of The Open University.

Printed in Great Britain by BPCC Wheatons Ltd, Exeter

CONTENTS

ABOUT THIS VOLUME   4

CHAPTER 1   **INTRODUCTION**

1.1   **MAPPING THE OCEANS**   5
1.1.1   Navigation   9
1.1.2   Depth measurement   11

1.2   **MAPPING THE OCEAN FLOORS**   13
1.2.1   Bathymetry from satellites   18

1.3   **UNDERWATER GEOLOGY**   21

1.4   **SUMMARY OF CHAPTER 1**   24

CHAPTER 2   **THE SHAPE OF OCEAN BASINS**

2.1   **THE MAIN FEATURES OF OCEAN BASINS**   27

2.2   **CONTINENTAL MARGINS**   29
2.2.1   Aseismic continental margins   29
2.2.2   Seismic continental margins and island arcs   31

2.3   **OCEAN RIDGES**   34
2.3.1   Ridge topography   34
2.3.2   Age-depth relationships across ridges   35

2.4   **TRANSFORM FAULTS AND FRACTURE ZONES**   36

2.5   **THE DEEP OCEAN FLOOR**   39
2.5.1   Abyssal plains   39
2.5.2   Seamounts   40
2.5.3   The distribution of submarine volcanoes   41
2.5.4   Aseismic ridges   43

2.6   **SATELLITE BATHYMETRY—A CASE STUDY**   43

2.7   **SUMMARY OF CHAPTER 2**   49

CHAPTER 3   **THE EVOLUTION OF OCEAN BASINS**

3.1   **THE EVOLUTION OF OCEAN BASINS**   52

3.2   **THE BIRTH OF AN OCEAN**   55
3.2.1   The Red Sea   56

3.3   **THE MAJOR OCEAN BASINS**   61
3.3.1   The Mediterranean   64

3.4   **SUMMARY OF CHAPTER 3**   65

CHAPTER 4	THE STRUCTURE AND FORMATION OF OCEANIC LITHOSPHERE	
4.1	**THE FORMATION OF OCEANIC LITHOSPHERE**	70
4.1.1	Pillow lavas: the top of the oceanic crust	70
4.1.2	Formation of the volcanic layer — a Case Study	72
4.1.3	Magma chambers at spreading axes	81
4.2	**SEGMENTATION OF OCEANIC SPREADING AXES**	82
4.2.1	A plausible model for lithospheric growth	84
4.2.2	Changes in spreading pattern	86
4.2.3	Crustal abnormalities	86
4.3	**SEAMOUNTS AND VOLCANIC ISLANDS**	88
4.4	**SUMMARY OF CHAPTER 4**	90

CHAPTER 5	HYDROTHERMAL CIRCULATION IN OCEANIC CRUST	
5.1	**THE NATURE OF HYDROTHERMAL CIRCULATION**	94
5.1.1	Heat flow, convection and permeability	94
5.2	**CHEMICAL CHANGES DURING HYDROTHERMAL CIRCULATION**	96
5.2.1	Changes in the rocks	97
5.2.2	Changes in seawater	99
5.3	**'BLACK SMOKERS'—AN EXERCISE IN PREDICTION**	101
5.3.1	'Black smokers', 'white smokers' and warm-water vents	102
5.4	**THE EXTENT OF HYDROTHERMAL CIRCULATION**	105
5.4.1	Variability in hydrothermal systems	107
5.4.2	Hydrothermal metamorphism	108
5.5	**MASS TRANSFER BY HYDROTHERMAL CIRCULATION**	109
5.6	**DISPERSAL OF DISSOLVED GASES AND OTHER HYDROTHERMAL EFFLUENT**	110
5.7	**SUMMARY OF CHAPTER 5**	111

CHAPTER 6	PALAEOCEANOGRAPHY AND SEA-LEVEL CHANGES	
6.1	**THE DISTRIBUTION OF SEDIMENTS**	114
6.1.1	Sediments and palaeoceanography	117
6.2	**CHANGES IN SEA-LEVEL**	120
6.2.1	Different time-scales in sea-level changes	121
6.2.2	The post-glacial rise in sea-level	122
6.2.3	Measuring Quaternary changes in sea-level	124
6.2.4	The growth of an ice-sheet: Antarctica	127
6.2.5	The salinity crisis in the Mediterranean	129
6.2.6	The migration of climatic belts	134
6.2.7	The effect of plate-tectonic processes on sea-level	135
6.2.8	Major transgressions and regressions	136
6.3	**SUMMARY OF CHAPTER 6**	137

CHAPTER 7	THE BROADER PICTURE	
7.1	**THE GLOBAL CYCLE**	139
7.1.1	Changes in components of the cycle	141
7.1.2	Some effects of short-term changes	143
7.1.3	The steady-state ocean	143
7.2	**SOME RATES COMPARED**	145
7.3	**SUMMARY OF CHAPTER 7**	146

APPENDIX	THE STRATIGRAPHIC COLUMN	148
	SUGGESTED FURTHER READING	149
	ANSWERS AND COMMENTS TO QUESTIONS	150
	ACKNOWLEDGEMENTS	166
	INDEX	168

ABOUT THIS VOLUME

This is one of a Series of Volumes on Oceanography. It is designed so that it can be read on its own, like any other textbook, or studied as part of S330 *Oceanography*, a third level course for Open University students. The science of oceanography as a whole is multidisciplinary. However, different aspects fall naturally within the scope of one or other of the major 'traditional' disciplines. Thus, you will get the most out of this Volume if you have some previous experience of studying geology, geochemistry or geophysics. Other Volumes in this Series lie more within the fields of physics, chemistry and biology (and their associated sub-branches).

Chapters 1 to 4 describe the processes that shape the ocean basins, determine the structure and composition of oceanic crust and control the major features of continental margins. Today's ocean basins are geologically ephemeral features, and these Chapters show why. Chapter 5 deals with the 'hot springs' of the deep oceans that result from the circulation of heated seawater through oceanic crust. This phenomenon was not even suspected until the mid-1960s and was not confirmed by observation until some years later. Since then, many people have seen the striking photographs of 'black smokers' at ocean ridges. Chapter 6 summarizes the main patterns of sediment distribution in the ocean basins and shows how sediments can preserve a record of past climatic and sea-level changes. Finally, Chapter 7 considers the role of the oceans as an integral part of global chemical cycles.

You will find questions designed to help you to develop arguments and/or test your own understanding as you read, with answers provided at the back of this Volume. Important technical terms are printed in **bold** type where they are first introduced or defined.

CHAPTER 1   INTRODUCTION

Humans have made use of the seas throughout recorded history and the unrecorded past for activities such as fishing, transport, trade and warfare. The first sea-going craft may have been developed from the river boats of the oldest known civilizations (Figure 1.1). Up to the fifteenth century there had been little attempt at systematic exploration of the oceans, though some epic voyages were made, especially in the Atlantic and Indian Oceans from Europe, North Africa and the Middle East, and over substantial areas of the western Pacific by Polynesians and Melanesians. Many regions were first explored by people other than Europeans. For example, in the early 1400s Chinese explorers reached the east coast of Africa by sea before the Portuguese, though their exploration was not followed by any attempt at colonization or trade.

Figure 1.1   A model of an ancient Egyptian river craft of about 2000 BC, from a tomb at Thebes.

1.1 MAPPING THE OCEANS

With the dawn of the Renaissance in the fifteenth century, Europeans began to dominate exploration of the oceans. Patrons such as Prince Henry the Navigator of Portugal encouraged such exploration as much for its own sake as to find new trade outlets. During the fifteenth to eighteenth centuries, Columbus, Magellan, Cook and others advanced seafaring techniques and our knowledge of global geography—and brought in their wake worldwide European colonization.

Early voyages of exploration were perilous, not least because of the difficulties of navigation and position-fixing out of sight of land. The few rudimentary instruments available to the navigator were limited to finding only the latitude of the ship at sea. No simple reliable means of finding the longitude was available until the development of accurate and reliable chronometers in the late eighteenth century.

Figure 1.2 Isogonic map by Edmund Halley, 1701, produced using data collected during his voyage on the *Paramour Pink*, 1698–1700. *Source*: N. J. W. Thrower (1972) *Maps and Man*, Prentice-Hall.

In view of these difficulties, Halley's map (Figure 1.2) is particularly remarkable. This shows that by 1701 the shape of the Atlantic Ocean basin had been determined, which was certainly not the case barely 200 years before (Figure 1.3). Halley's map shows coastlines quite accurately with respect to longitude, and also contours (isogons) of the amount of angular divergence between magnetic and true (or geographic) north.

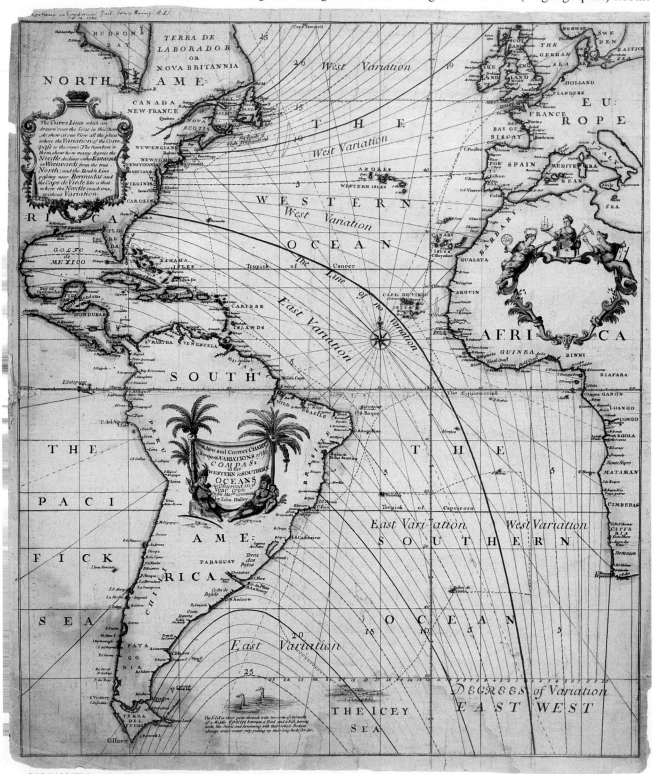

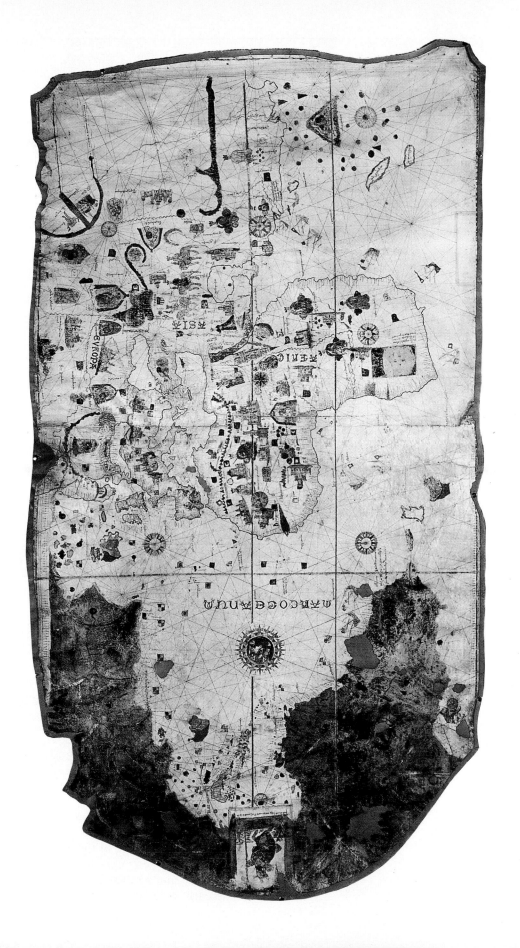

Figure 1.3 World map, c. 1500, by Juan de la Cosa, pilot on Columbus's second expedition. Courtesy of Naval Museum, Madrid; photographed by Arxiu Mas.

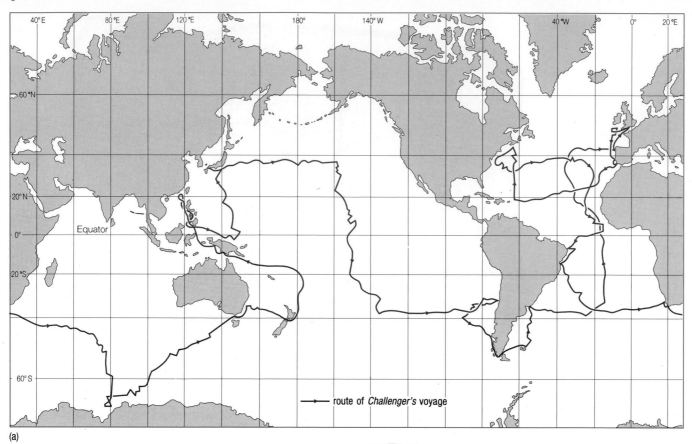

(a)

Figure 1.4(a)   The route of HMS *Challenger*, 1872–6.   (b)   HMS *Challenger*, 1872. She was a steam-assisted wooden corvette of 2306 tons.

(b)

The systematic gathering of oceanographic information was at first almost entirely concerned with geographical exploration and the preparation of coastal charts, so that knowledge of the oceans advanced little beyond the lore of seamen and fishermen. Very few investigations of the deep oceans were undertaken, but among the earliest was an unsuccessful attempt by Magellan in the early sixteenth century to sound the floor of the central

Pacific, which proved too deep for his sounding lines. These studies increased during the nineteenth century, culminating in the *Challenger* expedition of 1872–76, mounted by the British Government to collect oceanographic information (Figure 1.4).

The huge quantity of data of all types collected during *Challenger's* voyage was published in 50 volumes and may be considered to mark the birth of scientific study of the oceans. Other aspects of oceanographic research were also developed during this period. Until the nineteenth century, mariners were aware of ocean currents principally because their voyages were delayed or speeded up by them. For example, monsoonal reversals of the Somali Current, off East Africa, were documented as long ago as the ninth century. Sea-colour changes and the appearance of seaweeds far offshore (as in the Gulf Stream) provided further evidence of ocean currents. The difficulties of navigation in the open oceans meant that the velocity of ocean currents could not be related easily to fixed reference points in the way that the better known coastal currents could be related to land features.

1.1.1 NAVIGATION

Even though the problems of navigational accuracy experienced by the early explorers have been largely overcome by modern technology, it is still easier to navigate within sight of land than in the open oceans. Coastal navigation depends on the coast being visible from the ship. Factors limiting visibility (apart from bad weather and rough seas) are the Earth's curvature, the height of the observer (which may be a radar scanner) and the height of reference objects on land (Figure 1.5).

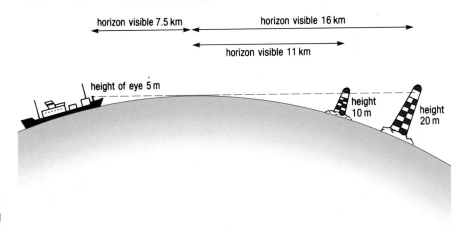

Figure 1.5 Visibility and the curvature of the Earth (not drawn to scale). An observer 5m above sea-level has a horizon at 7.5km, and can see objects 10m and 20m above sea-level at distances of 7.5 + 11 = 18.5km and 7.5 + 16 = 23.5km, respectively.

In bad weather or at night, radar (*radio detection and ranging*) is useful. Radio waves (at about 3000–10000 MHz[*]) are transmitted from the ship in a rotating beam and reflected from objects in its line of sight. The time between the transmission of the signal and reception of the reflection gives the distance, and the bearing is given by the direction from which the reflection arrives. The time between transmission and reception is very short, because radio waves are a form of electromagnetic radiation and travel at the speed of light ($\sim 3 \times 10^8 \, \text{m s}^{-1}$).

[*]MHz = megahertz. The hertz (Hz) is the unit of frequency = 1 cycle per second. 1 MHz = 10^6 Hz.

The accuracy of visual- and radar-fixing methods is of the order of tens to hundreds of metres, but they are limited to the line of sight and so have short ranges (Figure 1.5). Longer-range radio navigation systems use networks of land-based transmitters emitting pulsed or continuous radio signals with frequencies from 10kHz to 2MHz, which can travel round the curvature of the Earth. The ship's receiver measures the interference patterns set up between the radio waves from different stations, and the ship's position is found by reference to charts showing the positions of those patterns. The range and accuracy of such systems vary. For example, *Decca* has a range of about 500km, and a precision in optimum conditions of ±50m, whereas the *Omega* system has a range long enough to be used in the open oceans, with a precision of ±2–3km.

The classical method of navigation used in the open oceans is celestial navigation. The angles of the Sun, Moon, planets and stars above the horizon are measured using a sextant at a precisely recorded time. The motions of these bodies can be looked up in tables, and fixes on only two of these determine a unique location. Even under ideal conditions, celestial navigation is accurate only to within a few kilometres. Use of a sextant on a moving ship demands skill, and relatively laborious and time-consuming calculations are involved. Observations are possible only when both the horizon and the reference body are visible. Both the Sun and/or Moon and the horizon may be visible during daytime, but at night the horizon is not easily visible, so measurements using the planets or stars usually have to be made at twilight, when the horizon can be seen.

Celestial navigation was the mainstay of mariners for centuries, but now navigational satellites can provide positions to within ±100m anywhere in the open oceans, regardless of weather conditions. Satellite navigation was developed by the US Navy and came into general use in 1967. A complex network of navigational satellites is now available for mariners and aviators. Some satellites are in high orbits (about 20000km) and are 'visible' from a large part of the globe simultaneously. Others are in lower orbits, passing near the poles so that almost every point on the globe is covered as the Earth rotates (Figure 1.6). Ships and aircraft can receive position-fixing transmissions at least every few hours. Precise navigation is vital to many sorts of oceanographic study. As a matter of principle, more than one navigational method is commonly available in case of instrument failure. Although satellite position-finding is becoming the standard technique for research, commercial and military navigation, it seems unlikely that the older methods will ever be abandoned entirely. Table 1.1 summarizes the range and accuracy of the various navigational methods described in this Section.

Table 1.1   The range and accuracy of some navigational methods at sea.

Type	Availability and range	Accuracy
coastal (visual and radar)	line of sight (up to about 50km)	10–200m
radio navigation systems	variable, e.g. *Decca* up to about 500km;	10–100m
	Omega worldwide	1–15km
celestial navigation	worldwide	2–10km
satellite navigation	worldwide	0.1–200m

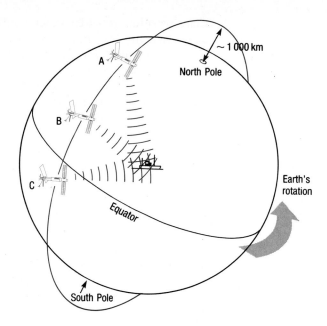

Figure 1.6 The principle of navigation by satellite. A satellite in polar orbit emits signals that carry information about the height of its orbit, its position and speed. While a satellite is passing (A, B and C represent successive positions), a shipboard computer analyses those data to determine its own position in relation to the satellite, and hence its position on the Earth's surface to within about 100 metres.

1.1.2 DEPTH MEASUREMENT

Contoured topographic maps of land areas are based on the measurement of height *above* a sea-level datum. In the oceans, contoured bathymetric maps are based on measurements of depth *below* a sea-level datum. Accurate positioning is of course essential for any height or depth measurement.

Before the development of echo-sounders, knowledge of ocean depths was limited to relatively few isolated soundings, which were often of doubtful accuracy. These soundings were made by lowering a weighted rope or steel plumb-line until the weight reached the sea-floor. The depth was assumed to be the length of cable let out. In shallow water, soundings could be made relatively rapidly (Figure 1.7), but in the deep oceans it would take several hours to lower and raise the sounding line. During this time, currents and winds could move the ship from the position where the weight was lowered, so the line ended up being far from vertical and the depth was often grossly overestimated. Furthermore, in deep water it was difficult to determine when the weight reached bottom, as the long length of cable was much heavier than the weight itself. An indication of the difficulties can be gauged from the fact that in their three-and-a-half-year expedition the *Challenger* scientists achieved only some 300 deep-water soundings.

Echo-sounders were first used on oceanographic expeditions in the 1920s. Hundreds of depth measurements could be made in a matter of days, and it became possible to produce accurate bathymetric maps of major features such as the Mid-Atlantic and Carlsberg Ridges, with their axial rift valleys.

Figure 1.7 The Leadsman. Line soundings were the only means of depth measurement until the early twentieth century. (*Artist:* Arthur Briscoe)

Echo-sounding provides rapid, continuous and accurate depth measurement. The ship transmits a sound pulse through the water, which is reflected from the sea-floor and received back at the ship. The interval between transmission and reception of the echo provides a measure of depth, using the known speed of sound in seawater. If pulses are transmitted at short intervals, reflections from the sea-floor can be recorded graphically to give a continuous profile of the sea-floor (Figure 1.8). Under ideal conditions, an accuracy of within a few metres is possible even in deep water. However, there are several sources of inaccuracy. The most obvious is that surface waves may move the ship up and down through several metres. More importantly, the speed of sound in seawater can vary by about ±4% according to variations in temperature, salinity and pressure. Oceanographers have tables for different areas of the world's oceans, which they use to make first order corrections of echo-sounder records (normally standardized to $1500\,\mathrm{m\,s^{-1}}$). Inaccuracies in these tables can lead to appreciable errors in depth measurements. For example, a 1% error in the sound velocity means a 30m depth error in 3km of water. Acoustic frequencies used in the ocean are generally about 5–30kHz.

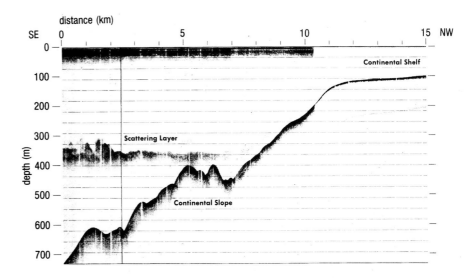

Figure 1.8   An echo-sounder record, giving a bathymetric profile of the edge of the eastern USA continental shelf. The slopes appear steep because the vertical exaggeration is ×12. The scattering layer probably represents a concentration of organisms and biological debris, and shows that echo-sounding has other applications.

Inaccuracies also result from the low resolving power of most echo-sounders. This is due to the divergence, at angles of about 30°, of the sound pulses as they are transmitted downwards, so that in deep water they cover large areas of ocean floor—about 1.5km diameter at 3km depth (Figure 1.9 (a)). Bathymetric variations over such an area will not be resolved, and the depth that is recorded could be shallower than the depth directly beneath the ship (Figure 1.9(b)).

Echo-sounders with higher resolution, using beam angles of a few degrees, have to be much bulkier in order to give a narrower beam, which is partly why they are not in general use. An equally important reason is that a large angle of divergence is necessary for routine navigational

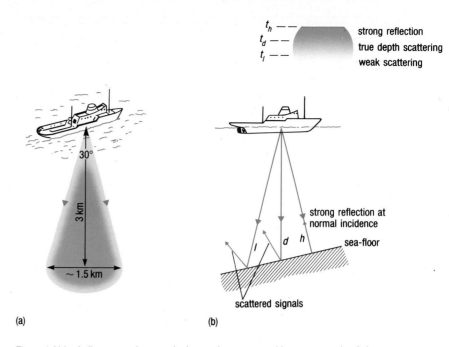

Figure 1.9(a)   A divergent echo-sounder beam gives a cone with an apex angle of about 30°.   (b)   Cross-section at an exaggerated scale to show how the true depth (d), directly below the ship, is masked by stronger reflections at normal incidence, giving a shallower depth value (h). $h < d < l$, so that times on the echo-sounder record (inset) $t_h < t_d < t_l$ show progressively longer travel times of sound pulses scattered and reflected from these different depths. This is a much simplified description but it illustrates the main problems, and shows that errors in depth-determination caused by poor resolution could be in the order of tens of metres in deep water.

purposes, to allow for use on a rolling ship in heavy seas. Narrow-beam instruments would be of little use in such situations, but are of course invaluable for survey work.

QUESTION 1.1   Examine Figure 1.8 and estimate the reliability of the data on position and depth shown on the profile for (a) the edge of the continental shelf, and (b) the lowermost 'bump' on the continental slope.

Question 1.1 illustrates the important point that in surveying the oceans it is often necessary to tolerate errors of greater magnitude than are normally acceptable on land, where mislocating certain topographic features with respect to the line of a proposed road by even a few metres, for example (let alone a couple of kilometres), would be unthinkable. However, positional accuracy needs to be very high indeed when siting a well-head on the sea-floor.

1.2   MAPPING THE OCEAN FLOORS

Knowledge of water depths in shallow seas has obviously always been important for sailors, and bathymetric charts for coastal waters have been produced for several centuries. The need for more extensive information about ocean bathymetry did not arise until relatively recently, and the first ocean-wide bathymetric chart was published by Matthew Maury in

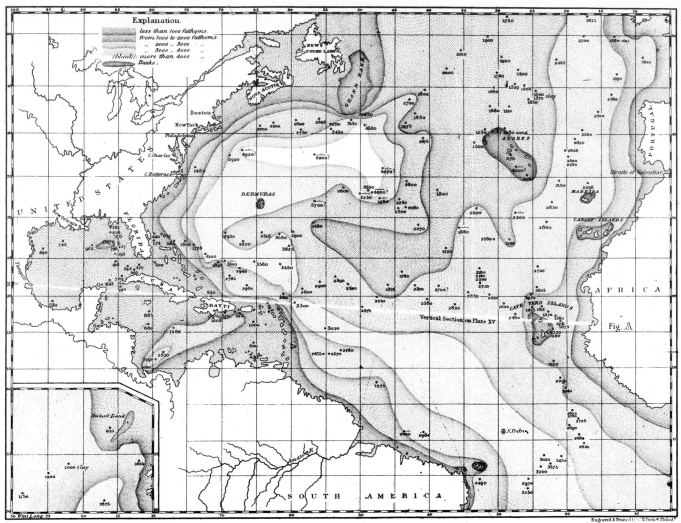

Figure 1.10(a)   Map of North Atlantic bathymetry by Matthew Maury, 1855. *Source*: M. Maury (1857) *The Physical Geography of the Sea*, Sampson Low & Son.

1855, for the North Atlantic (Figure 1.10). Maury's 'Middle Ground' was the first indication of the existence of the Mid-Atlantic Ridge, the long and sinuous form of which was soon revealed by the *Challenger* soundings, despite the limited number of measurements that could be made (Section 1.1.2).

The physiography of the ocean floor is now thought to be quite well known, but until recently even large features had not been recognized, and much detail still needs to be added. Thus, it was not until the 1960s that the International Indian Ocean Expedition delineated the Ninety-east Ridge (Figure 1.11) as a single 4500 km-long feature. In the Pacific Ocean, submerged seamounts and other features are still being discovered by satellites using altimeter measurements of the sea-surface (see Section 1.2.1).

Echo-sounding is a great advance on the early 'spot' soundings with lead lines, but what is the limitation of this method for ocean-floor mapping?

Echo-sounder records provide linear bathymetric profiles along the ship's track. Interpolation is necessary between profiles unless there are plenty of them along accurately known tracks that intersect one another at frequent intervals. This has happened only in very limited areas, which

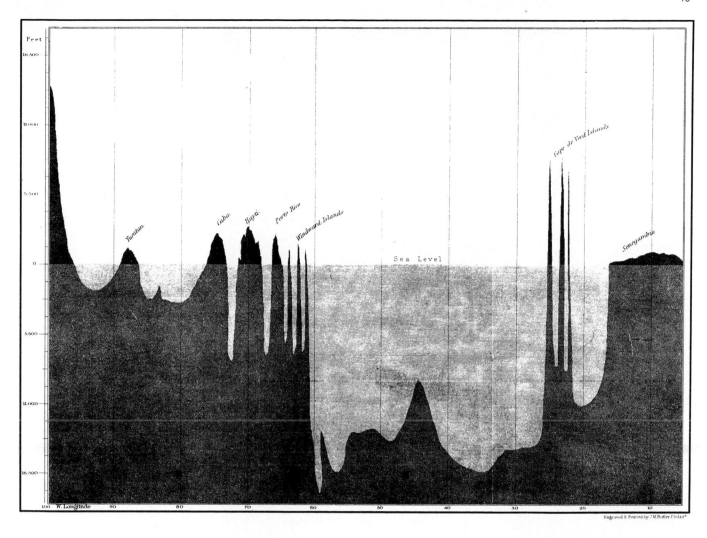

Figure 1.10(b) Profile of North Atlantic bathymetry by Matthew Maury, *op. cit.*

means that there are vast areas of ocean floor still relatively poorly mapped by this method.

The need for more extensive and precise bathymetric data continues to grow, for practical as well as academic reasons. Detailed maps of continental shelf and slope bathymetry are essential for exploration and exploitation of sea-bed mineral resources (e.g. petroleum, sands and gravels). It can also aid navigation, because echo-sounding can detect features such as submarine canyons, which in well-charted waters are sufficiently defined to permit accurate position-fixing (e.g. along the eastern US seaboard). In the open oceans, bathymetric information is essential for studies of deep circulation and the movement of sediments.

Since the 1960s, side-scan sonar systems have been developed, which are either stabilized hull-mounted instrument packages or housed in torpedo-like casings that can be towed behind ships at pre-determined depths. These send out fan-shaped signals that build up 'sound pictures' of the sea-floor on either side of the ship's track. One of the most sophisticated of these 'towfish' is *GLORIA* (*G*eological *LO*ng *R*ange *I*nclined *A*sdic). It scans the sea-floor with two divergent sonar beams that acoustically 'illuminate' the sea-floor in two swaths, each up to 30km wide, on either

16

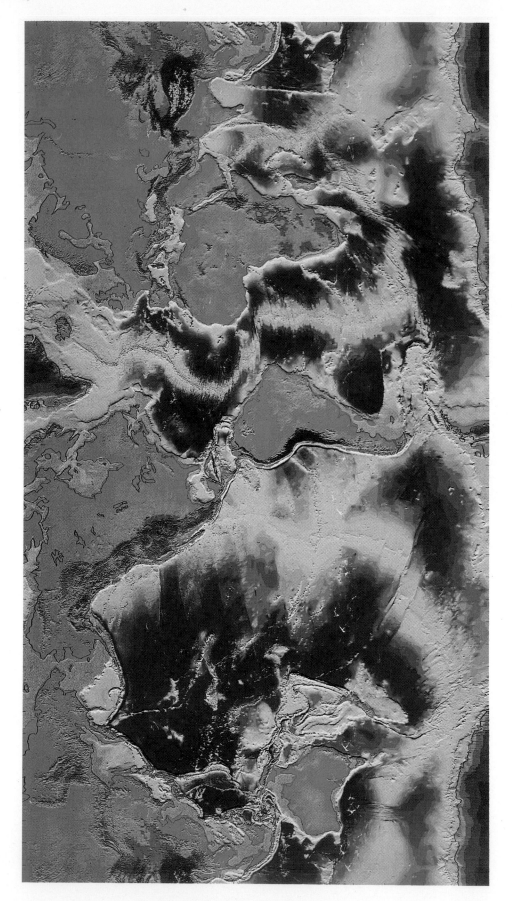

Figure 1.11   Shaded relief map of the Earth's solid surface. The Ninety-east Ridge is the linear ridge running almost north–south to the south-east of India near the right-hand edge of the map.

side of the ship's track (Figure 1.12). By overlapping the swaths so as to obtain a view of any object from both sides, ambiguities of interpretation are resolved.

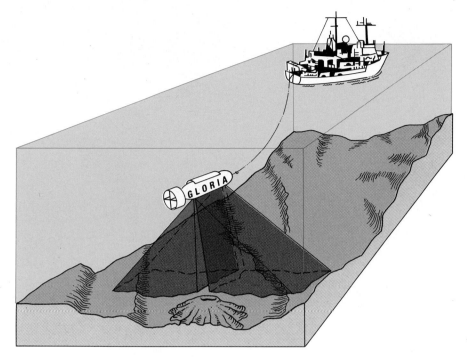

Figure 1.12 *GLORIA* is 8m in length and is towed 300m behind the mother ship at a speed of 10 knots and a depth of 50m, where it is neutrally buoyant—i.e. the overall density of the 'towfish' package is the same as that of the surrounding seawater. In water of 5000m depth, *GLORIA* can scan a swath of sea-floor about 60km wide and the time interval between sonar pulses (of 4s duration in the 6.2–6.8kHz range) is set at 40s to allow time for the most distant echoes to return. In water of shallower depths, the area covered and the time between pulses are reduced.

Sonographs produced by this technique not only reveal local topography, they can also indicate the nature of the bottom (Figure 1.13), information that is essential if breakages are to be avoided when laying pipelines or communication and power cables. Features which have been revealed by the operations of *GLORIA* and similar instruments (e.g. *Sea MARC*) include sinuous submarine canyons, submarine slides, hundreds of previously unknown submarine volcanoes, metal-rich deposits and manganese nodule fields.

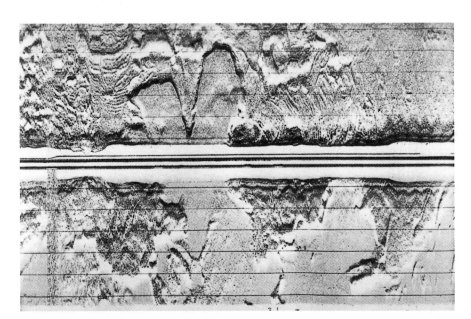

Figure 1.13 Part of a sonograph obtained from a side-scan sonar traverse at the entrance to the Black Sea. It shows bare rocky areas and sand-covered areas with sandwaves (megaripples). The central strip is the ship's path. The horizontal lines are 15m apart.

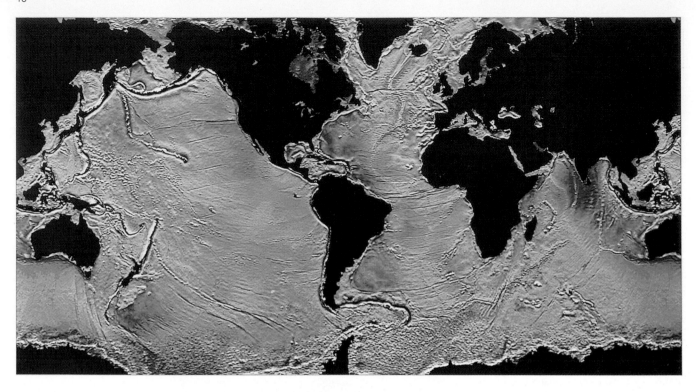

Figure 1.14 The marine geoid observed by the *Seasat* satellite. The data have been filtered to show detail but reduce the effect of very large-scale features.

1.2.1 BATHYMETRY FROM SATELLITES

In spite of the many depth measurements made since the 1920s, especially during and since the Second World War, some parts of the oceans remain poorly surveyed, notably the south-west Pacific and southern Indian Oceans. Side-scan sonar has improved detailed knowledge of many areas, but the number of instruments is small and they tend to be deployed for specific purposes in limited areas. Since the 1970s, altimeter measurements from satellites have helped considerably to refine available bathymetric maps of the oceans. We shall consider a specific example in Chapter 2, but here we focus on the principles involved.

Figure 1.14 shows the marine **geoid**, i.e. the shape of the surface of equal gravitational potential energy around the Earth, and—more importantly in

Figure 1.15 A satellite radar altimeter measures its own height above the sea-surface, which varies according to current velocities, wave action, tides, and barometric pressure. The sea-surface can be a metre or two *above* or *below* the surface of the geoid, which is mean sea-level in the absence of waves, tides, etc. The geoid surface is determined from the influence of spatial variations in gravity on the satellite orbit. It has vertical undulations giving a 'relief' of almost 200 m over the surface of the globe. Note that scales on this diagram are *greatly* exaggerated: geoid undulations (relative to a smooth reference spheroid) are two orders of magnitude greater than sea-surface topography.

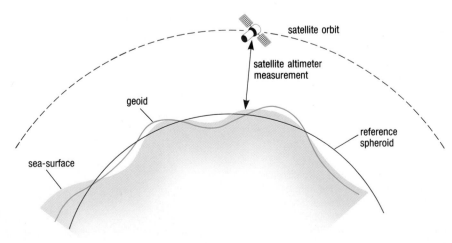

this context—the shape of mean sea-level in the absence of winds and currents. The geoid is not a smooth spheroid, because variations in the thickness and density of the Earth's crust and upper mantle cause perturbations in the gravity field. The geoid thus has highs and lows of variable wavelength, and its surface has a relief of up to 200m across the entire globe. In the oceans, undulations of the geoid correspond roughly with depth, so that sea-level is elevated over ocean ridges and depressed over trenches, though by only a few metres. Figure 1.15 illustrates the principles underlying the determination of bathymetry by satellite measurements of sea-surface elevation using highly accurate and precise radar altimetry. The relationship between altimeter measurements and the gravity field for part of the Mediterranean is illustrated in Figure 1.16.

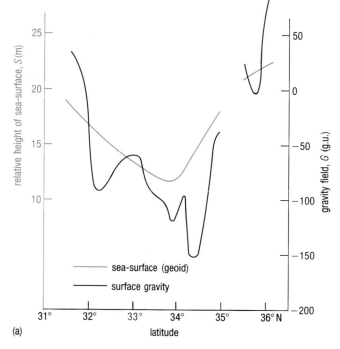

Figure 1.16(a)   An example from the eastern Mediterranean to show how variations in the Earth's gravity field (*G*), which were measured at sea-level, control the pattern of variation in mean sea-level (*S*), in the absence of waves, tides, currents and barometric pressure changes. *S* was measured by the altimeter aboard *Seasat*, and is the *geoid* surface. The gaps in the profiles correspond to the island of Crete. (g.u. = gravity units.)

(b)   Undulations in the mean sea-level surface (i.e. the geoid surface) of the Mediterranean, determined by the *Seasat* altimeter. The contour interval is 2m. Note the position of the profile, shown here in black, in part (a).

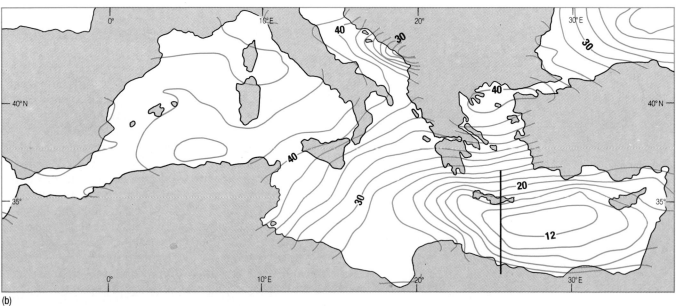

For smaller-scale bathymetric features in particular, there is an important correction to be made. Perhaps surprisingly, tides, currents and changes in atmospheric pressure can cause undulations of more than a metre in the ocean surface. Clearly, these effects must be filtered out if bathymetric details such as those depicted in Figure 1.17 are to be detected.

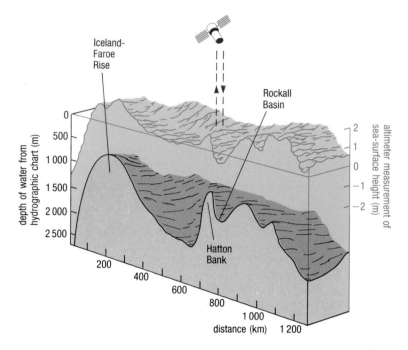

Figure 1.17   Block model of an area to the west of Scotland, based on *Seasat* altimeter measurements. Sea-surface elevation follows bottom topography to a remarkable degree. A sea-floor feature one or two thousand metres high can cause the sea-surface to hump up by a metre or two. *Important*: Note the very different vertical exaggerations for bathymetry (left-hand axis) and sea-surface elevation (right-hand axis).

QUESTION 1.2   Can you suggest a way of determining—and thus filtering out—these dynamic and therefore variable perturbations of the sea-surface due to tides, currents and climatic variations, when attempting to make corrections to bathymetric measurements obtained from satellite data?

Another remarkable outcome of the space programme was the discovery that radar data from oceanographic satellites such as *Seasat* could reveal the configuration of the sea-floor in shallow water. In this case the instrument used is not an altimeter, but is an imaging radar, which constructs an image of the sea- (or land) surface from a complex series of radar echoes recorded as the satellite passes by. This technique is analogous to side-scan sonar, and produces images such as that shown in Figure 1.18. However, the radar beam is incapable of penetrating more than about 1 cm below the sea-surface, so it is clear that the sand-banks and other bottom features are not being 'imaged' directly. What happens is that tidal currents flowing over an irregular sea-floor create a pattern of sea-surface roughness which is related to the bottom topography. These patterns are not in themselves a new discovery; indeed, they appear to have been known to the English captains blockading the Spanish Armada

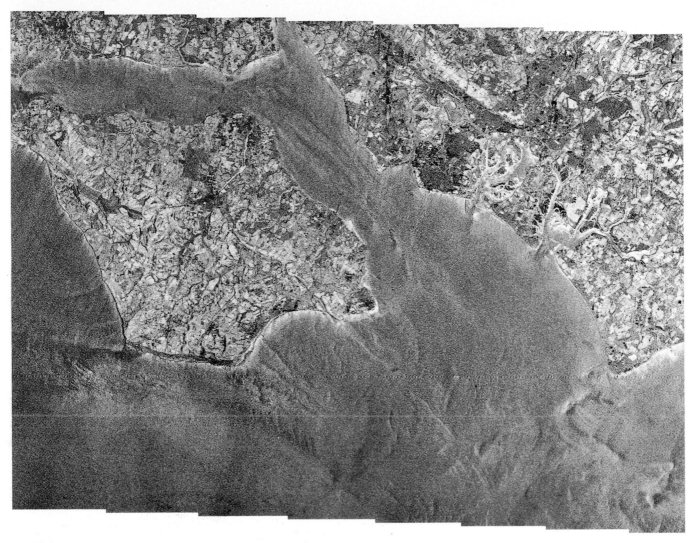

Figure 1.18 Patterns of sand-banks with sandwaves (megaripples) superimposed on them and other bathymetric features show up in this *Seasat* radar image of part of the English Channel.

in French ports in the sixteenth century. The real discovery was the close correlation between bottom features and sea-surface ripples of a wavelength that could interact with that of the radar beam (23.5 cm). Surveys of the sea-bed which otherwise could take weeks or months from a surface vessel can be achieved in a few seconds, if the satellite makes its pass at the right phase of the tidal cycle.

1.3 UNDERWATER GEOLOGY

Before 1930, virtually the only data on the nature and composition of the oceanic crust were from samples of rock and sediment dredged from the sea-floor by *Challenger* and later expeditions. Geophysical investigations in the decades immediately before and after the Second World War showed that the oceanic crust is denser than continental crust and has a distinctly layered structure. At the time, the consensus among geologists was still that the oceans and continents were fixed in their positions. Subsequent studies became caught up in the plate tectonics 'revolution' of the 1960s and early 1970s, which revealed just how fundamentally the oceans differ from the continents.

Figure 1.19 *Glomar Challenger*. Thrusters in the bow and stern were controlled by computer to maintain station within 30m of a point directly above an electronic beacon on the ocean floor. Drilling could thus proceed without snapping the drill string.

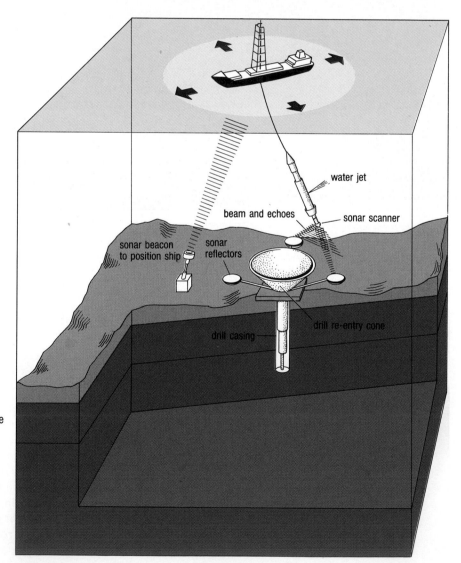

Figure 1.20 (Not to scale.) Re-entering a drill hole in the ocean floor: the re-entry cone is attached to the drill 'string' when it is first lowered to the bottom and it remains there when the string is withdrawn for the drill bit to be replaced. For re-entry, the drill string is lowered again, with a sonar scanner mounted on the bit assembly. This emits sound signals that are echoed back from three reflectors spaced round the cone. Information about the position of the bit assembly relative to these reflectors is used in the control of water jets which steer the bit directly over the cone.

The Deep Sea Drilling Project (DSDP) was begun in the 1960s at the start of this 'revolution', using the *Glomar Challenger* (Figure 1.19), a ship that was specially equipped for drilling into the ocean floor beneath several kilometres of seawater. The main technical advance which made this possible was the development of a re-entry system whereby the drill bits could be replaced and the drill string repositioned in the hole (Figure 1.20). Many hundreds of holes have now been drilled through the ocean-floor sediments into the underlying igneous rocks, and have added greatly to our understanding of crustal structure and sediment distribution. The *Glomar Challenger* was succeeded by a similar but better-equipped vessel, the *JOIDES Resolution*, which began operations in the international Ocean Drilling Project (ODP) in 1985. One of her early achievements was to sink the first drill hole into rocks beneath a zone of hot springs (hydrothermal vents) at an ocean ridge crest, penetrating layers of copper, iron and zinc sulphides.

Examination of the oceanic crust has been further facilitated by the use of instruments lowered down drill holes to measure the physical and chemical properties of the sediments and underlying volcanic rocks *in situ*. The drilling programme has been supplemented by increasingly sophisticated echo-sounding techniques, side-scan sonar and underwater photography from cameras towed close to the sea-floor, which have enabled very accurate bathymetric maps and profiles to be compiled.

Since 1973, geologists have been able to do 'field-work' under several kilometres of seawater using manned submersibles (Figure 1.21), equipped with precise navigation systems (including sonar), sampling apparatus and cameras. The use of remote-controlled submersibles achieved great publicity in 1986 when one took video pictures inside the wreck of the *Titanic*. Such vehicles typically carry cameras, side-scan sonar equipment, and sampling gear, and can reduce the time taken for

Figure 1.21 The submersible *Alvin* at the start of a dive. First used on project FAMOUS (see Chapter 4), *Alvin* has since been used to investigate many key areas of the ocean floor.

deep-sea exploration and survey from years to weeks, because they can dive deeper and stay down longer than manned submersibles. They are also cheaper and involve virtually no risk to human life.

Our knowledge of the form, structure and evolution of ocean basins is growing rapidly, and we hope to give you some flavour of these developments in the following Chapters.

1.4 SUMMARY OF CHAPTER 1

1   Accurate navigation at sea became possible only with the development of accurate and reliable chronometers in the eighteenth century, which enabled longitude to be determined. Nowadays, there are many radio-navigational aids, both surface and satellite-mounted.

2   Accurate depth determinations became possible only with the development of echo-sounding early in this century. This is still the main technique for bathymetric measurements. Side-scan sonar instruments allow substantial areas of sea-floor to be surveyed at once.

3   Satellite altimetry uses radar to measure very accurately the height of the sea-surface. After corrections for transient effects (waves, tides, currents and atmospheric pressure changes) have been made, the mean sea-level can be determined. This follows a surface of equal gravity—the geoid. Irregularities in the geoid correspond with topographic features on the ocean floor, and the height of the sea-surface can therefore provide a measure of bathymetry. Imaging radar can show up bathymetric features in shallow water.

4   Detailed exploration of oceanic crustal structure and composition began in the 1960s, with the DSDP. Since then, much has been learned, not least through the aid of new technologies for imaging and through submersible operations, both of which supplement standard acoustic navigational and surveying equipment.

Now try the following question to consolidate your understanding of this Chapter.

QUESTION 1.3   Estimate the approximate wavelength and amplitude of geoid undulations in the area depicted in Figure 1.17.

CHAPTER 2   THE SHAPE OF OCEAN BASINS

Few students of modern Earth sciences have not heard of sea-floor spreading and plate tectonics, or have failed to see maps and diagrams such as Figures 2.1 and 2.2.

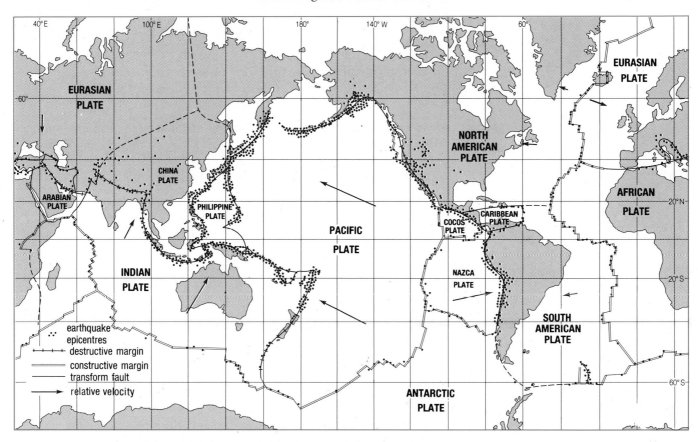

Figure 2.1   The world pattern of plates, ocean ridges, trenches and transform faults in relation to earthquake epicentres indicated by red dots. Tentative positions of plate margins are indicated by dashed lines. There are seven major plates (capital letters) and six minor ones (small capital letters), plus several smaller ones, not named here. The length and direction of the arrows indicate the relative velocities of the plates, averaged over the past few Ma. The African plate is assumed to be stationary. The arrow shown in the key corresponds to a relative velocity of 5 cmyr^{-1}.

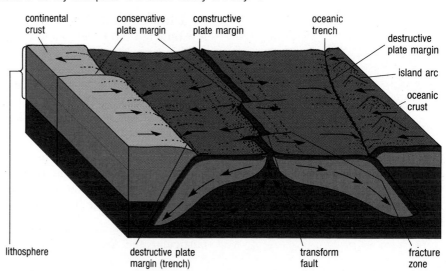

Figure 2.2   Vertically exaggerated diagram showing the basic concepts of plate tectonics. Plates of rigid lithosphere (which include oceanic or continental crust and uppermost mantle) mostly between about 100 and 250km thick overlie a layer of relatively low strength called the asthenosphere. Mantle material rises at constructive plate margins (ocean ridges or spreading axes) and plate material descends into the mantle at destructive plate margins (ocean trenches). At conservative plate margins, plates merely slide past each other.

We provide below a numbered summary of the main features relating to these processes, partly by way of a reminder and partly to provide a basis for what follows in the rest of this Volume.

1   The rigid, outer shell of the Earth—the **lithosphere**—is up to 250 km thick under the continents but is only about 100 km thick under the oceans. The lithosphere consists largely of **peridotite**, a dense rock which forms the upper part of the Earth's **mantle**. The uppermost part of the lithosphere is the **crust**. Below the lithosphere is a much weaker layer within the mantle called the **asthenosphere**, which allows the lithospheric plates to move.

2   Beneath continents, the top 35 km of the lithosphere, on average, is formed of continental crust. This may reach 90 km in thickness beneath mountain ranges. Continental crust is mostly granitic in composition.

3   Oceanic crust is both thinner (7–8 km) and denser than continental crust. It is basaltic in composition. (In this Chapter, references to oceanic crust relate to the igneous crust formed by magmatic processes at spreading axes; we do not usually include the sediments as well.)

4   Oceanic crust lies below sea-level whereas continental crust is mostly much higher. This is a result of **isostasy**: the Earth's outer layer tends towards gravitational equilibrium by height adjustments between areas of different relative buoyancy.

What topographic features on Figure 1.11 correspond to the **constructive** and **destructive plate margins** illustrated in Figures 2.1 and 2.2?

They are described below in items 5 and 6.

5   The global system of constructive margins or sea-floor **spreading axes** forms the nearly continuous ridge system that resembles a mountain chain snaking its way through the major ocean basins. This is where oceanic lithosphere is generated.

6   The numerous ocean trenches and island arcs—most of them within the Pacific—are sites of oceanic plate **subduction** and re-absorption into the mantle. The Alpine–Himalayan mountain chain bears witness to a phase of subduction and continental plate collision that is geologically quite recent (within the past 150 Ma).

7   The polarity of the Earth's magnetic field has reversed at intervals of a few hundred thousand to a few million years (Ma), during the past 100 million years or so at least (before that, the intervals appear more variable). The alternations of 'normally' and 'reversely' magnetized 'stripes' of ocean floor lying parallel with the constructive margins at which they formed (Figure 2.3) has enabled the age of the ocean crust to be mapped and sea-floor spreading histories to be compared among the different ocean basins.

8   There appears to be no oceanic crust older than about 160 Ma in any of the major oceans, so the present-day ocean basins are very young in relation to the age of the Earth (4600 Ma). Furthermore, they are changing all the time, in two major respects:

(a)   Their shape and size change because of plate movements. In particular, the Atlantic and Indian Oceans must be expanding, because

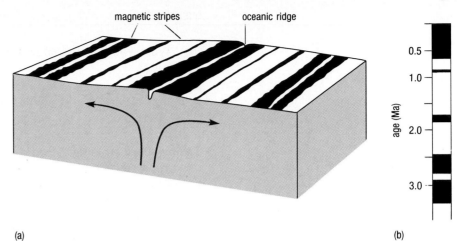

(a)

(b)

Figure 2.3(a) Asthenospheric mantle material rises under an oceanic ridge to produce new oceanic lithosphere. As this moves sideways, it cools past the Curie point of magnetic minerals in the rocks and so 'freezes in' the polarity of the Earth's field that prevails at the time. (b) A simplified polarity time-scale for the past few million years, showing alternations of normal (black) and reversed (white) polarity of the Earth's magnetic field.

they have spreading axes but no major subduction zones, but the Pacific is contracting because subduction there outpaces spreading. The rates of change are geologically quite large; Figure 2.1 shows that plates move at time-averaged speeds of up to several cm per year.

(b)   Their bathymetry changes because of (i) contraction as rock masses cool (see Section 2.3.2); (ii) deposition of continent-derived sediments; (iii) erosion and submarine canyon formation due to currents; and (iv) isostatic adjustments.

2.1 THE MAIN FEATURES OF OCEAN BASINS

Figure 2.4 shows that nearly half of the Earth's solid surface lies within two quite well-defined limits of altitude (0–1 km) and depth (4–5 km), and that a significant proportion lies within a few hundred metres of sea-level.

Figure 2.4   The distribution of levels on the Earth's surface.

(a)   A histogram showing the actual frequency distribution.

(b)   The hypsographic curve: a cumulative frequency curve based on (a). This is *NOT* a profile of the Earth's surface; it is a curve showing the percentages of the Earth's surface that lie above, below, or between any given levels.

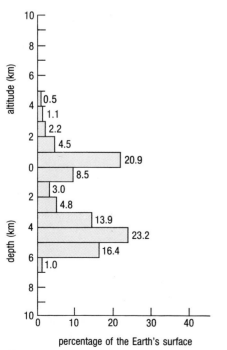

(a)

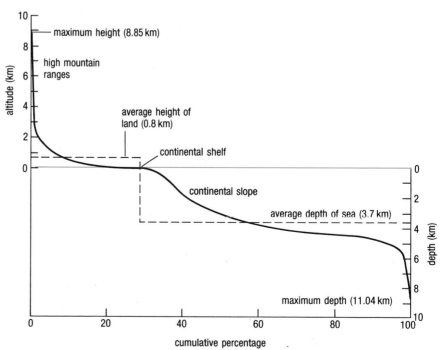

(b)

QUESTION 2.1 (a) What percentage of the Earth's surface lies below sea-level?

(b) What would be the effect, on this percentage, of a 100m rise in sea-level?

(c) Taking the mean radius of the Earth as 6370km, what percentage of this radius is represented by the total vertical range in relief of the Earth's solid surface?

The two most striking features of the ocean floor are the world-wide ridge-rift systems traversing all the major ocean basins (and in places encroaching on the continents) and the circum-oceanic (especially circum-Pacific) systems of deep trenches. New oceanic lithosphere is continually being formed along the ridges (or spreading axes) by sea-floor spreading. The lithosphere so formed eventually descends back into the asthenosphere at the trenches which are the surface expression of the inclined planes known as **subduction zones**.

In between the ridges and trenches lie sediment-covered abyssal plains, as well as subsidiary ridges, and a variety of hills and plateaux, some of which break the surface as islands. Bordering the continents are the continental shelves, formed of thick accumulations of sediment. Here, water depths are generally 200m or less, and the width of the shelf varies greatly from place to place.

The relative importance of the main features responsible for ocean bathymetry is summarized in Table 2.1. Figure 2.5 is a semi-diagrammatic cross-section—with greatly exaggerated vertical scale—across the South Atlantic and South America (*cf.* Figure 1.11), to illustrate the main physiographic features listed in Table 2.1. In the following Sections, we discuss the various components of the ocean basins in more detail, beginning with the edges of the continents.

Table 2.1 Main features of the principal ocean basins.

	Ocean			
	Pacific	Atlantic	Indian	World ocean
ocean area (10^6km²)	180	107	74	361
land area drained (10^6km²)	19	69	13	101
ocean area/drainage area	9.5	1.6	5.7	3.6
average depth (m)	3940	3310	3840	3730
area as % of total:				
shelf and slope	13.1	19.4	9.1	15.3
continental rise	2.7	8.5	5.7	5.3
deep ocean floor	42.9	38.1	49.3	41.9
volcanoes and volcanic ridges*	2.5	2.1	5.4	3.1
ridges†	35.9	31.2	30.2	32.7
trenches	2.9	0.7	0.3	1.7

*Volcanic ridges are those related to volcanic island chains that are not part of constructive plate margins, e.g. the Walvis Ridge in the South Atlantic. They do not include island arcs.

†These are the ridges that correspond to constructive plate margins, e.g. the Mid-Atlantic Ridge.

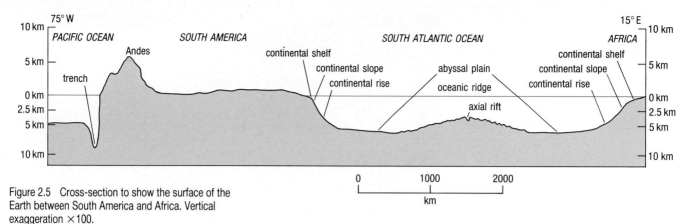

Figure 2.5 Cross-section to show the surface of the Earth between South America and Africa. Vertical exaggeration ×100.

2.2 CONTINENTAL MARGINS

Two basic types of continental margin had been recognized even prior to the advent of plate tectonics. These are Atlantic and Pacific types.

Atlantic-type margins usually have a relatively wide continental shelf and an extensive continental rise (Figure 2.5 and Table 2.1). Because they have relatively little earthquake activity, they are termed **aseismic** or **passive margins**. They develop when continents are rifted apart and form new oceans. The continent and adjacent ocean floor are part of the same plate. Areas of continental crust may become completely isolated due to rifting, and form **microcontinents**. These may be almost entirely below sea-level (in which case the crust is anomalously thin, like the Rockall Bank and Seychelles Plateau), or may be quite large island masses such as Madagascar.

Pacific-type margins typically have a trench at the foot of the continental slope, replacing the continental rise (Figure 2.5). Table 2.1 shows that rises cover much smaller areas and trenches much greater areas in the Pacific than in the other oceans. Pacific-type margins are now more commonly known as **seismic** or **active margins**, because they are seismically very active (i.e. earthquakes occur frequently), forming where an oceanic plate is being consumed beneath a continental plate at a subduction zone. Here the continent and ocean floor belong to different plates. Seismic margins, with associated island arcs, can also form at the boundary of two oceanic plates—a special case that we shall consider later. Continental margins round the Indian Ocean are mainly of Atlantic type, except in the vicinity of the Java Trench, in the north-east (Figures 1.11 and 2.1).

2.2.1 ASEISMIC CONTINENTAL MARGINS

Aseismic margins develop by crustal stretching and eventual rifting apart of a continent and are modified by subsequent sediment deposition. A cross-section through such a margin would look something like Figure 2.6. In plan, the shape of an aseismic margin is related to the trend of the original rift. In cross-section, the shape is determined partly by the amount of stretching, faulting (rifting) and associated crustal subsidence, and partly by the extent of sediment deposition which forms the shelf–slope–rise zone. All of these vary considerably from place to place,

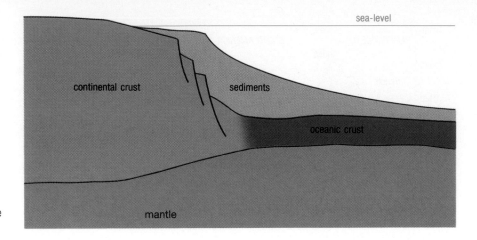

sea-level

continental crust   sediments

oceanic crust

mantle

Figure 2.6 Highly diagrammatic cross-section with great vertical exaggeration and not to scale, through an aseismic continental margin. Details of the surface profile are described in Figure 2.7 and related text.

so that although the basic features can generally be discerned, aseismic margins may differ enormously in detail.

QUESTION 2.2 (a) Continental crust is typically thinner than average at aseismic continental margins. Why should thinned (stretched) continental crust subside?

(b) Looking at Figure 2.7, approximately where would you place the transition between continental and oceanic crust?

Figure 2.6 shows that the sediments forming most of the continental shelf must rest mainly on thinned continental crust (*cf.* Question 2.2(b)). The width of the continental shelf can be as much as 1500 km, and the surface is generally flat, with an average gradient of only 0.1° (Figure 2.7). However, in places there may be transient undulations due to sand-waves and banks up to a few metres high, resulting from the action of currents.

Water depth at the edge of the continental shelf—the shelf break, see Figure 2.7—varies from 20 to 500 m, but averages around 130 m. The continental slope has a width of between 20 and 100 km, and its base is between 1.5 and 3.5 km in depth. The slope has a much higher gradient than the shelf, averaging about 4°, which is very steep for submarine relief where gradients are usually low compared with the land. However, aseismic margins formed by recent rifting of a continent may have continental slopes much steeper than this because of the angle of the initial rift. The Gulf of California, for example, which began opening only about 4 Ma ago, has slopes steeper than 20° because these have not yet been modified by erosion and sedimentation, processes which have been active along the Atlantic's ocean margins for over 100 Ma.

In many areas, such as the Western Approaches (south and west of the British Isles), the continental slope is cut by **submarine canyons**. These act as channels for the transport of sediment to the deep ocean. Most submarine canyons start on the continental shelf, commonly (though not invariably) at the mouths of major rivers. They are generally V-shaped and resemble river valleys on land; but they are due to the erosive action of **turbidity currents**—sediment and water mixtures that are denser than seawater and flow even down gentle slopes, reaching velocities high enough to scour out the canyons.

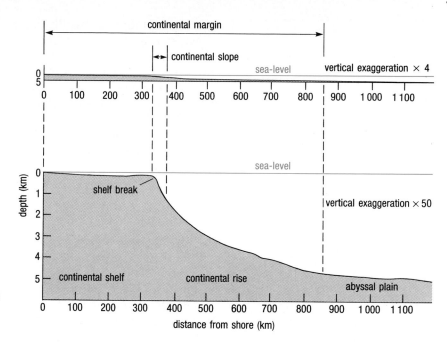

Figure 2.7 One possible configuration of an aseismic continental margin, showing the continental rise in relation to shelf and slope. The transition from continental slope to rise should be at about 2 km depth.

Once the turbidity currents reach the foot of the continental slope, they slow down and begin to deposit their loads, building up the sediment wedges which form the continental rise. This has a much gentler gradient than the slope, about 1° on average. It may be up to 600 km wide, depending partly on the strength and frequency of turbidity currents and partly on the erosive power of currents in the deep ocean circulation. The surface of the continental rise may also be cut by channels in which turbidity currents continue to flow across the rise to the abyssal plains (Figure 2.7).

2.2.2 SEISMIC CONTINENTAL MARGINS AND ISLAND ARCS

Seismic continental margins are usually associated with an oceanic trench, which marks the site of subduction of the oceanic lithosphere into the asthenosphere. They are destructive plate margins.

Trenches may occur next to:

1 continental margins with volcanic coastal mountain ranges, where the oceanic lithosphere is subducted beneath a continent.

2 island arcs, where the oceanic lithosphere is subducted beneath another oceanic plate.

From what you have read so far, what do you think is likely to happen to the sediments carried by turbidity currents flowing down seismic margins, and how do you think the continental slopes of seismic margins differ from those of aseismic margins?

Sediments carried by turbidity currents are generally trapped by trenches at the foot of the continental slopes, which tend to be steeper than those of aseismic margins (hence the small area of continental rise in the Pacific—see Table 2.1).

These contrasts are well displayed in the long Peru–Chile trench system, which marks the subduction zone for two oceanic plates, the Nazca Plate

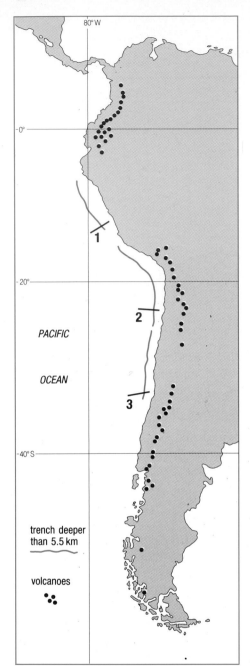

Figure 2.8 The western margin of South America, showing segments of the Peru–Chile Trench that are deeper than 5.5km. The red dots are active volcanoes. Lines 1–3 correspond to the bathymetric profiles in Figure 2.9.

and part of the Antarctic Plate, beneath the western coast of South America (Figure 2.1). A high range of mountains—the Andes—extends along the length of the coast. Associated with the subduction zone are intense seismicity and volcanism and a characteristic gravity anomaly pattern (negative at the trench, but positive over the volcanic mountain arc) that characterizes destructive plate margins. Figure 2.8 shows that the Peru-Chile Trench is not continuous. Gaps in oceanic trenches are not uncommon and are believed by some authorities to be due to the subduction of seamounts or small aseismic ridges (Sections 2.5.2, 2.5.4).

QUESTION 2.3   Examine Figures 2.8 and 2.9.

(a)   How do (i) the width of the continental shelf and (ii) the shape of the continental slope compare with those typical of aseismic margins?

(b)   Is the gradient of the continental slope in general consistently greater than anything you might find on aseismic margins?

In some cases (e.g. profiles 1 and 2 in Figure 2.9), differences in the depth and width of the trench are possibly linked with the rate of subduction. For example, faster subduction could result in a deeper and narrower trench. However, there can be other causes of differences in trench profile.

To what would you attribute the differences in the actual trench region illustrated in profiles 2 and 3 of Figure 2.9?

The flat floor of the trench shown in profile 3 must be due to sediment filling the trench. This in turn prompts a further question, namely: why do all the profiles not show flat floors underlain by thick sediments, which the high relief and close proximity of the Andes would be expected to provide?

At least so far as profiles 2 and 3 are concerned, one possible explanation lies in present climatic conditions:

The Atacama desert in northern Chile has less than 0.01m of average annual rainfall, so there is virtually no fluvial transport of sediment to the ocean—and the trench in profile 2 reaches a depth of more than 8km. Further south, however, annual rainfall rises to over 4m, enabling large amounts of sediment to be transported to the trench, completely filling it south of about 50°S.

Sediments within trenches generally show evidence of deformation, resulting from active tectonic movement. Extensional (or normal) faulting is characteristic of the outer, oceanic, face of the trench wall, caused by warping of the oceanic plate as it bends downwards into the subduction zone. Inner trench walls often carry highly folded and thrust-faulted (reverse-faulted) sediments, which have been scraped off the descending oceanic plate so that the inner wall builds out progressively oceanwards. These features are illustrated in Figure 2.10, a profile obtained by seismic reflection.

Only some of the sediment in the trench is incorporated on the inner trench wall. The remainder is carried down into the mantle by the descending plate. The relative proportions of material scraped off and subducted vary considerably from place to place. It is the subduction and

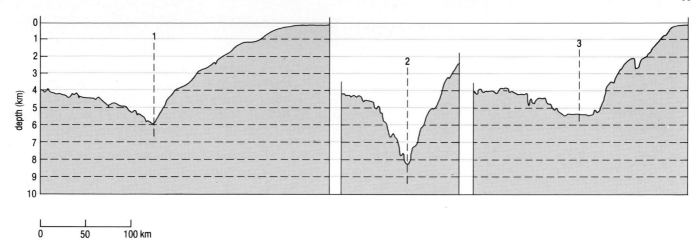

Figure 2.9 Bathymetric profiles across the Peru–Chile Trench at locations given in Figure 2.8. The vertical exaggeration is ×25.

subsequent partial melting of the oceanic plate and its sediment load that is responsible for the build up of a volcanic mountain chain (like the Andes) on the continental plate, parallel to, and usually no more than a couple of hundred kilometres from, the plate margin.

QUESTION 2.4 If you look at Figure 2.10 you will see that the ocean-floor sediments are scraped off as slices or wedges. Would you expect the oldest-formed wedges to be at the top or bottom of the sequence forming the inner trench wall?

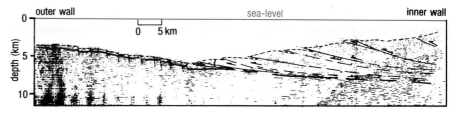

Figure 2.10 A seismic reflection profile with line interpretation across the Middle America Trench (just north of the Peru–Chile Trench), showing extensional faulting downwards into the trench on the outer wall, and thrust-faulting on the inner wall. The vertical exaggeration is ×1.5.

Island arcs are similar to Andean-type mountain chains, but occur when one oceanic plate is subducted beneath another, instead of beneath a continent. The same sort of trench occurs when the two plates meet, and a volcanic island arc develops on the overriding plate, above the subduction zone. Interestingly, a small ocean basin usually opens on the far side of the island arc from the trench. This is called a **marginal basin** or **back-arc basin**. The occurrence of sea-floor spreading in what might simplistically be expected to be a compressional environment, because it is where two plates are converging, indicates that plate motions are associated with a complex interplay of forces and magmatic activity and are not completely understood. In parts of the western Pacific several sets of island arcs and marginal basins have formed (*cf.* Figure 1.11), by progressive splitting apart of older arcs.

We now leave the continental margins and subduction zones and move into the deep oceans, to examine the other end of the sea-floor spreading conveyor belt, where oceanic lithosphere is generated.

2.3 OCEAN RIDGES

These are the most important physiographic features of the ocean basins as well as being the sites of formation of new oceanic lithosphere—the constructive margins. Depending on how their edges are defined, ocean ridges currently occupy about 33% of the total area of ocean floor (Table 2.1) as well as a large volume of the ocean basins (see Figure 2.5).

Basaltic magma is extruded and intruded along the median zone of ridges, a process accompanied by the separation of the flanking lithospheric plates, so that new ocean floor is being continually formed. The rate of sea-floor spreading is usually more or less symmetrical about the ridge, adding to the plates on either side at the same average rate, and the direction of spreading is usually approximately perpendicular to the ridge.

2.3.1 RIDGE TOPOGRAPHY

The physiographic characteristics of ocean ridges are related to the spreading rate. Representative topographic profiles across a slow-spreading ridge (the Mid-Atlantic Ridge, 1–2 cm yr^{-1}*) and a fast-spreading ridge (the East Pacific Rise, 6–8 cm yr^{-1}*) are shown in Figure 2.11.

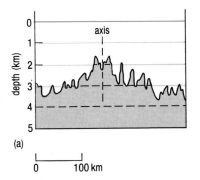

(a)

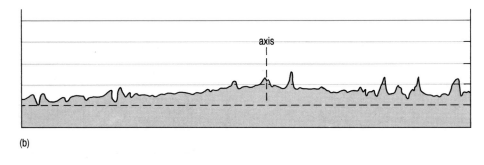

(b)

Figure 2.11 Representative east–west topographic (bathymetric) profiles across the Mid-Atlantic Ridge and across the East Pacific Rise (see Question 2.5). The vertical exaggeration is ×50.

QUESTION 2.5 The main features of these two ridges are summarized below. Read this summary, refer to the profiles in Figure 2.11, and then answer (a)–(d).

The Mid-Atlantic Ridge has a median valley about 25–30 km wide and 1–2 km deep; this is not so well developed and may even be absent on the East Pacific Rise. Gradients away from the ridge crest are of the order of 1 in 100 for the Mid-Atlantic Ridge, closer to 1 in 500 for the East Pacific Rise.

(a) Which profile relates to the Mid-Atlantic Ridge and which to the East Pacific Rise?

(b) Which ridge has the rougher surface relief?

(c) How do the average gradients of ocean-ridge flanks compare with those of continental margins?

(d) The Carlsberg Ridge in the northern Indian Ocean has an average spreading rate of 1–2 cm yr^{-1}. Which profile in Figure 2.11 should it more closely resemble?

*Unless stated otherwise, spreading rates are half-rates, i.e. the rate at which a plate moves away from the spreading ridge rather than away from the other plate.

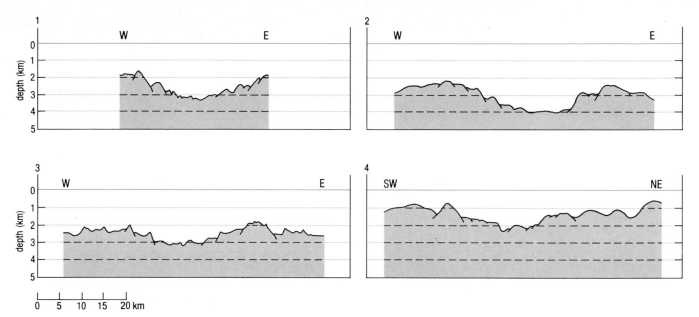

Figure 2.12 Detailed topographic profiles across the median valleys of: the Mid-Atlantic Ridge at 47° and 22°N (Profiles 1 and 2); the Gorda Ridge, north-east Pacific Ocean (3), and the Carlsberg Ridge, north-west Indian Ocean (4). The vertical exaggeration is ×5 throughout.

Profiles across the median valleys of three ocean ridges (Figure 2.12) show that these are rift valleys formed by normal faulting on fault planes inclined towards the axis. On the flanks beyond the valleys there are some outward displacements, along faults inclined away from the axis (e.g. the right-hand end of profiles 2 and 3). This block faulting is responsible for much of the rough topography of ridge crests.

2.3.2 AGE–DEPTH RELATIONSHIPS ACROSS RIDGES

One of the most interesting features of ocean ridges is that the depth of water increases systematically with the age of the lithosphere (Figure 2.13). This results in water depth increasing with increasing distance from the ridge. The reason for this relationship is that the lithosphere is hottest near the ridge axis, where it is forming, and is therefore least dense and isostatically most 'buoyant'. As the lithosphere cools over millions of years while moving away from the ridge, it contracts, becomes less buoyant and subsides. The age–depth relationship can be used to

Figure 2.13 Observed and theoretical relationship between the depth to the top of the oceanic crust and its age. The solid line is a best-fit curve through observed points. The dashed line is a theoretical elevation curve, calculated on the assumption that an increase of depth with age is due to the thermal contraction of the lithosphere as the plate cools on moving away from the ridge axis. Magnetic anomaly numbers refer to the linear magnetic stripes on the ocean floor, which are arranged symmetrically about ridge axes.

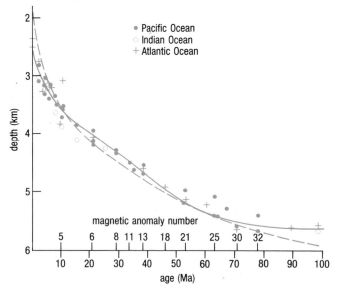

determine the approximate age of oceanic crust of known depth, and can help in the construction of palaeobathymetric (ancient bathymetry) maps which show how ocean basins developed.

There are some limitations, however:

1 The age–depth relationship cannot be used for oceanic lithosphere older than about 100 Ma, as by then the lithosphere has lost most of its heat of formation and approached thermal equilibrium, i.e. the curve has levelled off.

2 The depth of the ridge axis itself varies with the rate of spreading (fast-spreading ridge axes usually being deeper than slow ones); the average depth to the axis of the East Pacific Rise is about 2.7 km, that to the Mid-Atlantic Ridge axis is about 2.5 km.

2.4 TRANSFORM FAULTS AND FRACTURE ZONES

Ridge axes may be offset along **transform faults** (Figure 1.11), where oceanic plates slip sideways past each other. The transform faults lie along arcs of small circles about the pole of relative rotation between pairs of plates, and they are parallel to the direction of relative movement of the plates (Figure 2.14). Transform faults are so called because they link two points at the ends of successive ridge segments where spreading motion is transformed into sideways slip.

Because of the relative motions of the plates on either side of them, transform faults are seismically active. It is important to realize that

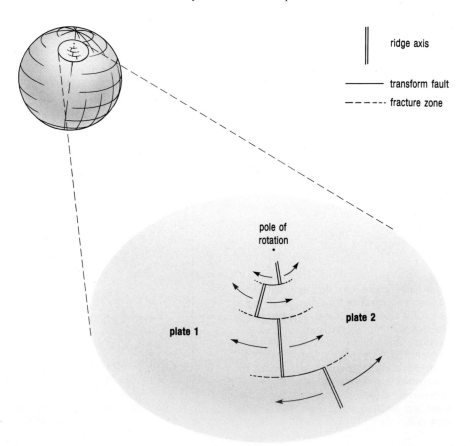

Figure 2.14 Both transform faults (heavy lines) and their inactive extensions, fracture zones (dashed lines) are small circles centred on the pole of relative rotation of the two plates. The spreading rate is related to the angular velocity and the distance from the rotation pole. Spreading rates therefore increase gradually with distance from the rotation pole, as indicated by different lengths of arrows.

transform faults occur only at plate boundaries and that they finish where they meet the ends of the two offset ridge segments which they connect. Beyond those points, the extension of a transform fault lies within a single oceanic plate, has no relative lateral motion across it and shows no major seismic activity. Such a feature is called a **fracture zone** (Figure 2.14).

If the original rift between two continents is at an oblique angle to the spreading direction, the spreading axis will usually develop as a series of offset ridge segments, as illustrated in Figure 2.15. Weak zones in rifting continents probably control the sites where the more important transform faults develop, as may have happened when South America and Africa split apart to form the equatorial Atlantic Ocean.

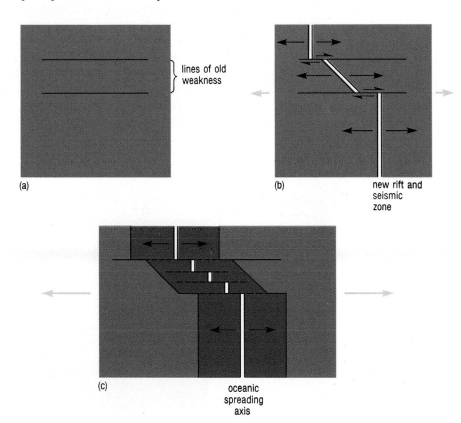

Figure 2.15   The break up of a continent, with lines of ancient weakness controlling the development of major transform faults, which offset the ridge segments and become aseismic fracture zones outside the zone of active spreading. Ridges are normally perpendicular to spreading directions, though not invariably so.

Figure 2.16   The elevation of the sea-floor is different on opposite sides of transform faults and fracture zones, so that escarpments are formed.

Transform faults and fracture zones are usually expressed as bathymetric features such as scarps and clefts on the ocean floor. The age–depth relationship described in Section 2.3.2 explains the origin of scarps (Figure 2.16): on opposite sides of a transform fault or fracture zone the lithosphere is of different ages and therefore lies at different depths. Fracture zones are not totally aseismic, because adjustments in the relative depth of ocean floor on either side of a fracture zone must occur as the lithosphere cools, contracts and subsides at different rates on either side. Generally, this produces only small earthquakes. Vertical displacements accompanied by minor earthquakes can occur along fracture zones near continental margins if different thicknesses of sediment accumulate on either side of a fracture zone because of differences in the sediment load brought in by rivers. This can lead to differential loading, and greater subsidence on one side of the fracture than the other.

Transform faults differ fundamentally from the transcurrent faults found on land: the actual movement along transform faults is in the *opposite* direction to that suggested by the ridge offsets, because of spreading at the ridge axis (Figures 2.14–2.16).

Large transform faults are also known as **conservative plate margins** (e.g. the San Andreas Fault in western North America and the New Zealand Alpine Fault, Figure 2.1), because there is normally no component of subduction or spreading along them. One of the best-known exceptions is the plate boundary in the Atlantic Ocean between the European and African plates (Figure 2.1). This is a major transform fault, but it is not a wholly conservative plate margin. At its western end, near the Mid-Atlantic Ridge, it has a small spreading component as well as the dominant transcurrent (lateral) motion, and there is some leakage of magma from the upper mantle, leading to eruption of basaltic lava along it. This type of fault is known as a 'leaky transform'. Further east, nearer to Gibraltar, it is a typical conservative margin, and the only movement is transcurrent. Where it enters the Mediterranean it acquires a significant component of subduction, again superimposed on the dominant lateral movement. Smaller transform faults can also show these features, especially a small spreading component, and leaky transform faults are not uncommon along ocean ridges (Figure 2.17).

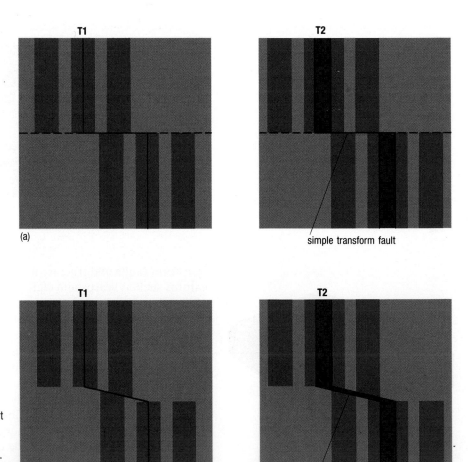

Figure 2.17 If a transform fault occurs at right angles to the ridge segments which it joins (a), then it lies parallel to the direction of sea-floor spreading, and spreading progresses (T1 to T2) without generation of new sea-floor along the transform fault. However, if a transform fault occurs oblique to the direction of spreading (b), then the fault has to be 'leaky', to fill the space left as the plates diverge.

Transform faults and fracture zones are simple in principle, but complex in detail, as revealed by surveys using side-scan sonar equipment (e.g. *GLORIA*, *Sea MARC*, Section 1.2) and submersibles, in addition to geophysical measurements. For example, displacement may occur either along a single main fault, or along a complex of branching smaller faults. In some areas, particularly on fast-spreading ridges such as the East Pacific Rise, offsets of about 10 km or less between segments of spreading axis can be sustained without the development of a transform fault. Instead, the two ridge segments propagate parallel to one another and begin to overlap. Eventually, one segment curves into the other, extinguishing and cutting off a portion of the old axis. These overlapping spreading centres are described at greater length in Chapter 4.

2.5 THE DEEP OCEAN FLOOR

Between the continental margins and the ocean ridges lies the deep ocean floor, representing about 42% of the total oceanic area (Table 2.1). This is a region of varied topography (bathymetry) and relief (Figure 1.11), comprising featureless abyssal plains, ridges, seamounts, plateaux or banks (which may be microcontinents or products of submarine volcanism), and long narrow clefts (the surface expression of major fracture zones).

2.5.1 ABYSSAL PLAINS

Abyssal plains have gradients of less than 0.05°; a change in height of less than 1 m per km. This is flatter than any other ocean feature and much flatter than most land areas. The flatness is interrupted in places by abyssal hills and seamounts, which often rise abruptly from the plain, suggesting that their bases are buried by sediment. Abyssal hills are not more than a few hundred metres high, whereas **seamounts** are defined as exceeding a kilometre in height (Section 2.5.2). Both are probably volcanic in origin.

Characteristically, there is a change of gradient between the foot of the continental rise and the abyssal plain. The gradient of the plain decreases away from the continental rise until it reaches the foothills of the ridge, known as the abyssal hill province, where the fault-controlled topography of the ridge is only partly buried by sediment.

QUESTION 2.6   Look at Figure 2.18, which summarizes the description we have just given. From which major ocean basin might this profile have come?

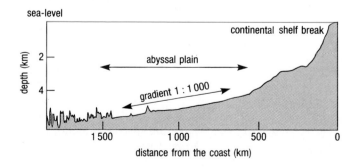

Figure 2.18   A topographic profile across an abyssal plain and continental rise and slope.

depth (km)

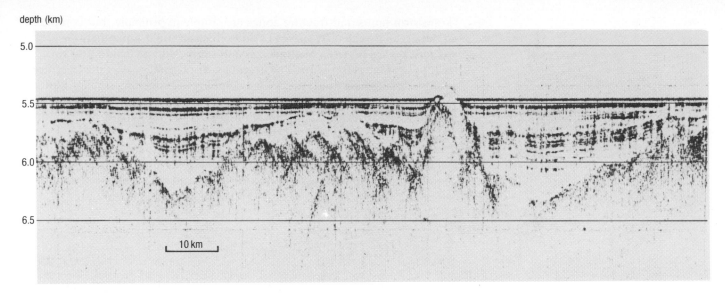

Figure 2.19   A seismic reflection profile across part of the Madeira abyssal plain (north-east Atlantic). The vertical exaggeration is ×20.

Turbidity currents deposit most of their load to build up the sediment wedge of the continental rise, but some of the load is carried further out and, along with **pelagic sediments** (deposited from suspension in the open oceans), gradually blankets the rough topography of the oceanic crust that is formed at ridge axes. Figure 2.19 shows the rugged topography that can underlie abyssal plains, and how effective the burial process is, with only the tops of hills remaining exposed on the surface.

Would you expect abyssal plains to be as extensive in the Pacific Ocean as in the Atlantic?

The Pacific Ocean is bordered either by trenches or by marginal seas behind island arcs, both of which trap almost all the sediment brought in from continental areas by turbidity currents. Although there are several abyssal plains in the Pacific, their sediment cover is predominantly pelagic, rather than continental, in origin.

2.5.2   SEAMOUNTS

The majority of seamounts form as ocean-floor volcanoes, most of which do not rise above sea-level. Those that do are called oceanic islands, but the term seamount is commonly used to describe both islands and subsea volcanoes. Seamounts are found throughout the oceans, but are particularly common in the Pacific, partly because of their less complete burial by abyssal plain sediments.

Figure 2.20 shows the topographic profiles of some seamounts. The angles of the slopes can be as great as 25°, so seamounts are among the steepest features on the ocean floor. Seamounts are roughly circular in plan and some are very large; the base of Mauna Loa (which forms the largest of the Hawaiian Islands) is about 100km across and its summit is about 9km above the surrounding floor, making it about as high above its base as Mount Everest.

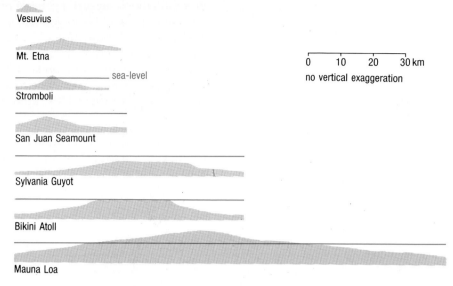

Figure 2.20   Topographic profiles across some on-land volcanoes and seamounts (including guyots).

Many seamounts are of similar age to the surrounding ocean floor, and must have formed on or near a spreading ridge axis. Some once reached above sea-level, but are now below the surface. This is partly because of the general subsidence of the whole oceanic lithosphere as it moves away from a spreading ridge, and partly because of more local subsidence, in isostatic response to the additional load of the seamount itself on the crust. Flat-topped seamounts, or **guyots** (Figure 2.20), are generally believed to have been planed off by wave erosion when the volcanoes were at sea-level. (But see also Chapter 4.)

The tops of many guyots are well below sea-level, far beyond the reach of the action of waves or surface currents. Would you expect to find a relationship between the age of the sea-floor and the depth to the top of guyots?

In general, we would expect the summits of older guyots to be at greater depths and further from ridge axes, having subsided more since their formation (*cf.* Figure 2.13). In equatorial regions, many subsiding seamounts and guyots have been massively colonized by corals, coralline algae and other wave-resistant shelly organisms so that carbonate reefs and atolls have been built up, the tops of which are maintained near sea-level.

Where seamounts have formed by volcanism away from the spreading axis, on older oceanic crust which had cooled down since its formation, the surrounding crust has been reheated, giving rise to a thermal anomaly. Such crust is warmer and thus more buoyant than expected for its age, and plots above the age–depth curve of Figure 2.13.

2.5.3   THE DISTRIBUTION OF SUBMARINE VOLCANOES

Many submarine volcanoes appear to be randomly scattered about the ocean floors (Figure 1.11). Others, however, are arranged in linear chains. Figure 2.21 shows four linear chains of volcanic islands and seamounts in the Pacific Ocean, all aligned roughly NW–SE. Age determinations suggest a general progression in the ages of individual volcanoes from older to younger south-eastwards along each chain. This has led to the formulation of the '**hot spot**' concept, according to which each linear chain is formed as a result of the Pacific Plate moving over a narrow heat source, or **mantle plume**. This consists of pulses of material rising from deep in the mantle to a point fixed relative to the Earth as a whole, and across which the lithospheric plate moves (Figure 2.22).

There is a regular progression of ages along the Hawaiian–Emperor Chain, and Figure 2.21 shows that the age range along the Hawaiian Chain is 43 Ma over a distance of 3400 km.

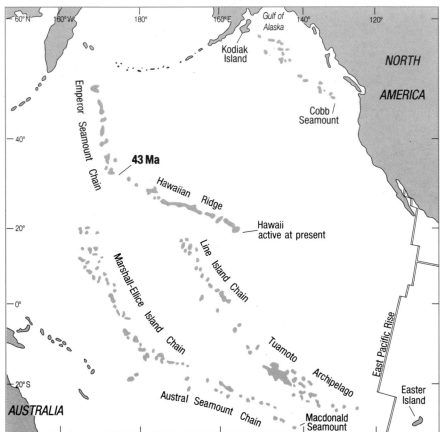

Figure 2.21 Four seamount and island chains in the Pacific Ocean. The youngest volcanoes are at the south-eastern end of each chain, and the age of the seamount at the bend in the Hawaiian–Emperor Chain is shown.

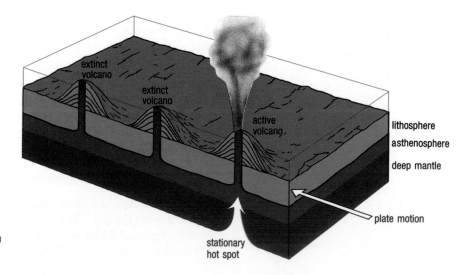

Figure 2.22 Schematic diagram (not to scale) illustrating how a volcanic island chain could be formed by an oceanic plate moving over a stationary hot spot or mantle plume. The age of the islands increases towards the left. New islands will appear on the right as the motion continues.

QUESTION 2.7 (a) What would be the average rate of movement of the Pacific Plate according to the hot-spot model?

(b) Given that a constant rate of movement was maintained, approximately what would be the age of the northern end of the 1900km-long Emperor Chain?

(c) Suggest a reason for the kink in the chain as a whole (at the western end of the Hawaiian segment, the southern end of the Emperor segment).

(d) In fact, the age of the northern end of the Emperor Chain has been dated at about 72Ma. Assuming that a change in *rate* of plate motion coincided with the change in *direction* of motion you have just postulated, what was the average rate of movement of the Pacific Plate between 72 and 43Ma ago?

Data from the Hawaiian–Emperor Chain provide strong support for the hot-spot model; but evidence from other chains is more equivocal. For example, although there is an overall south-eastward decrease in age within the nearby Line Islands–Tuamoto Chain, it contains many seamounts with anomalous ages. There is no simple explanation. Some workers have proposed the existence of more than one hot spot to account for the complications within this chain. Others appeal to evidence for a widespread episode of off-ridge volcanism, unrelated to hot spots, throughout the Pacific about 80–90Ma ago.

2.5.4 ASEISMIC RIDGES

These are fairly prominent and more or less continuous features traversing the deep ocean floor, and rising in some cases to more than 3000m above it. Examples include the Walvis and Rio Grande Rises in the South Atlantic, and the remarkably long and straight Ninety-east Ridge in the Indian Ocean (Figure 1.11). A hot-spot mechanism has been proposed as the origin for these features, and the available evidence supports this suggestion.

QUESTION 2.8 What must be the main difference in the character of the volcanism forming aseismic ridges from that forming a seamount/island chain?

2.6 SATELLITE BATHYMETRY—A CASE STUDY

This Section outlines how *Seasat* radar altimeter measurements were used to refine bathymetric maps in parts of the relatively less well-surveyed southern oceans. In Section 1.2.1 you saw that the geoid (the surface of gravitational equipotential) round the Earth and thus also the mean sea-level in the absence of winds, tides and currents, has a surface relief of as much as 200m. This relief is the result of the effects both of large-scale and small-scale gravitational variations. Thus, there are long wavelength ($c.500$km) undulations in the geoid which result from deep-seated mass anomalies within the Earth, and much shorter wavelength undulations reflecting shallow structures and sea-floor topography. The magnitude of longer wavelength effects is large enough to obscure the shorter wavelength features, so the longer wavelength trends must be measured and removed (or simply filtered out) from the altimeter data obtained from the satellite before the bathymetric information becomes apparent.

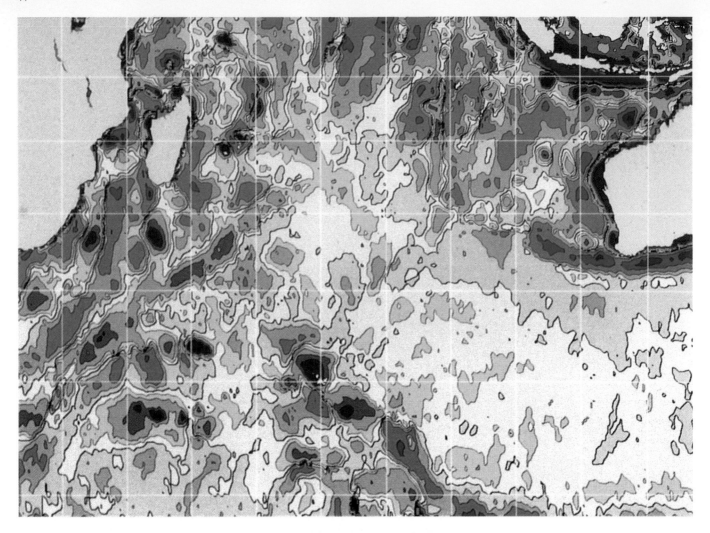

Figure 2.23(a)   Geoid anomaly map of the South Indian Ocean derived from *Seasat* mean surface measurements, processed to remove larger-scale anomalies (more than *c.* 500km). The contour interval is 1m, except for the +1 to −1 range, where it is 0.5m. Darkest blue represents areas deeper than −4m, brightest red represents areas higher than +4m. The zero contour (heavy black line) is between lightest blue and palest pink.

Seasat data were collected over 70 days during the 100-day lifetime of the satellite from 5 July to 10 October 1978. Temporal variations related to transient oceanographic effects were minimized because of the repeated traverses over the same points during different orbits. The *Seasat* altimeter was sufficiently precise to define the distance from the satellite to the sea-surface to within 1 part in 10^7, or about 5cm, but the height of the satellite in its orbit was known only to within about 1.6m. However, such errors in satellite position appear mostly as long wavelength variations, and do not have a large effect on the short wavelength features relevant to bathymetry. By the use of averaging techniques, *Seasat* data can produce a map of the marine geoid with a horizontal resolution of about 50km and an internal precision in height of about 20cm.

Figure 2.23 covers two areas in the southern oceans. The data were processed to remove the longer-wavelength deep-seated effects. This left only the effects of shallow structures and variations in sea-floor topography, producing geoid anomaly maps (Figure 2.23(a) and (c)). The maps in Figure 2.23 (b) and (d) are drawn from previous bathymetric information obtained from shipborne observations.

These pictures illuminate two important ways in which satellite information is of great assistance in ocean exploration.

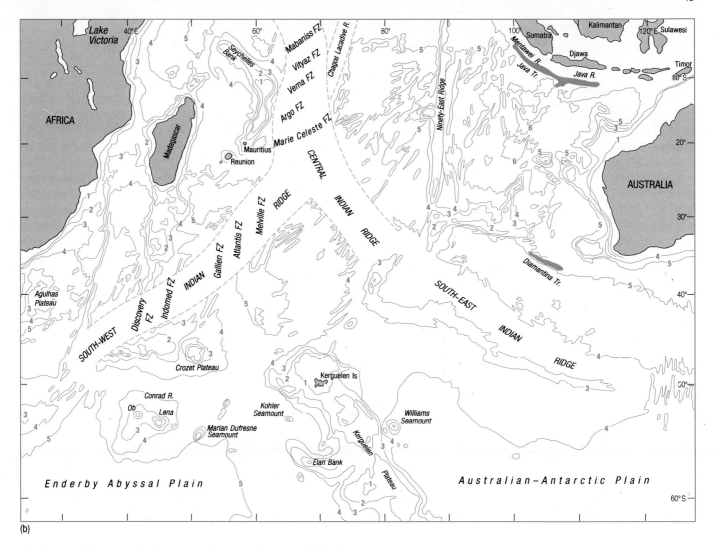

(b)

(b) Bathymetric map, compiled from conventional shipborne measurements of the South Indian Ocean. Area north-east of Java Trench not contoured. Complex topography less than 4 km depth on active ridges (between dashed lines) not contoured. Contour interval = 1 km; diagonal lines are trenches; R = ridge or rise; FZ = fracture zone; Tr = trench.

Improved bathymetry

Previously unidentified features are located, and the bathymetry of known features can be refined. This is especially well exemplified in Figure 2.23(a) for the Indian Ocean, particularly in the vicinity of the Kerguelen Plateau, which is poorly known, chiefly because ship tracks in this region are sparse or non-existent. By way of example, two of the newly distinguished features on the geoid anomaly map are:

1   A large trough east of the Kerguelen Plateau, between 54°S and 60°S.

2   A small plateau between Crozet and Kerguelen Islands, centred on 45°S, 56°E.

Seasat also provided significant new bathymetric information in the Pacific. Several major seamounts on the Louisville Ridge (Figure 2.23(d)) are apparent in the *Seasat* data (c) but do not appear in bathymetric compilations, because they are in regions never covered by ship tracks. Two examples occur near 163°W, 42°S, and 169°W, 37°S, and there is a very large positive feature at 153°W, 48°S. The north-west portion of the Louisville Ridge appears to be a closely spaced series of linear segments, each *c.*300 km long and consisting of two or three large, closely spaced seamounts. These features are consistent with a volcanic, hot spot, origin for the ridge.

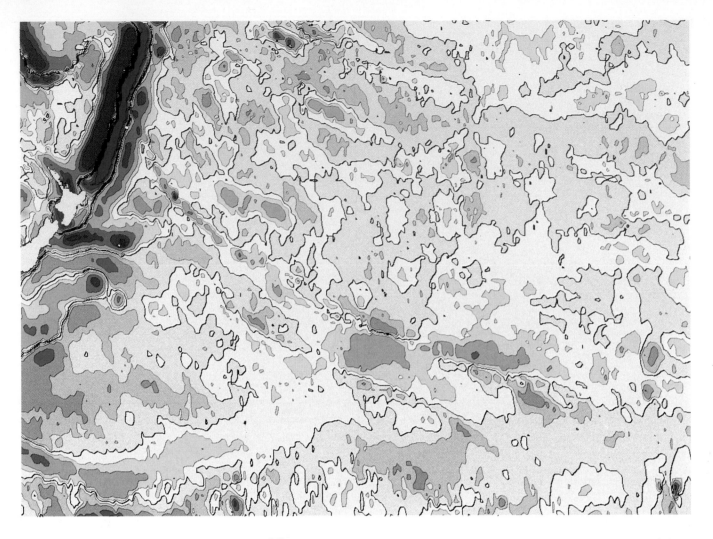

(c)   Geoid anomaly map of the South Pacific Ocean derived from *Seasat* mean surface measurements, as in (a).

QUESTION 2.9   (a)   On Figure 2.23(a), what features show up west of the Ninety-east Ridge, and how well do they correlate with the known bathymetry?

(b)   On Figure 2.23(b), how does the bathymetric expression of the Diamantina Trough relate to its expression on the geoid anomaly map (a)?

(c)   How good is the correspondence between the geoid map and the bathymetric chart for the three plateau regions in the western Indian Ocean: (i) Agulhas Plateau, (ii) Madagascar Plateau (south of the island of Madagascar) and (iii) Mozambique Plateau (between the two, nearer the coast of southern Africa)? What is the size of the geoid anomaly over these features?

(d)   In general, how do fracture zones show up on the geoid anomaly maps?

(e)   How are the Tonga–Kermadec Trench (north of New Zealand) and the sea-floor on either side of it expressed on the geoid anomaly map (Figure 2.23(c))?

In Question 2.9(e) you encountered one example of how bathymetry inferred from geoid anomalies alone must be treated with caution. There are two other possible sources of significant error. First, there may be

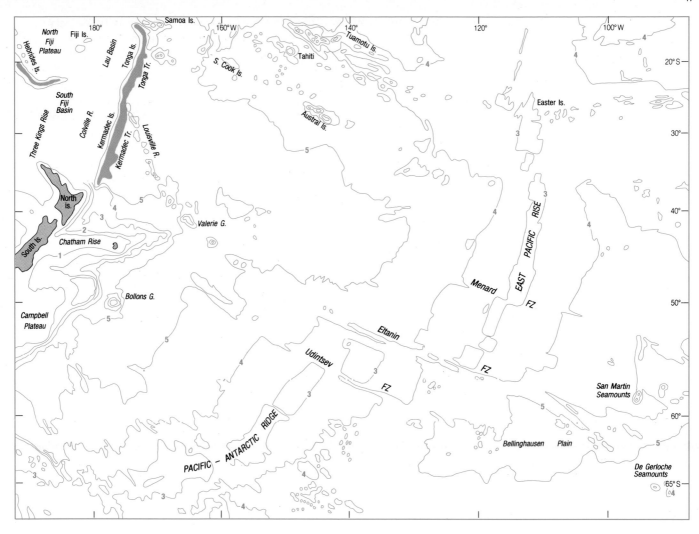

(d) Bathymetric map, compiled from conventional shipborne measurements as in (b), of the South Pacific Ocean. Complex topography west of Tonga–Kermadec Trench not contoured.

some bathymetric features without significant geoid anomalies over them. This is because as the plates cool, they thicken and become more rigid. For example, seamounts which formed on young oceanic crust, when the flexural rigidity was low, will have a relatively low geoid response, because the load has been locally compensated (Figure 2.24(a)). The load due to features formed on older more rigid crust will be spread out over wider areas of the plate (Figure 2.24(b)), and its effect on the geoid will be greater.

Figure 2.24(a) A younger and more flexible plate will accommodate the load of a seamount locally. The gravitational anomaly (and hence the geoid anomaly) over it is less than in (b) which shows a seamount formed on an older, more rigid plate.

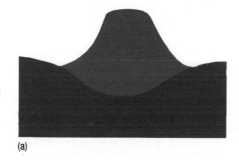

The second problem is the converse of the first: there may be density anomalies in the crust and upper mantle which have no surface expression, or at any rate less than might be expected. A possible example is on Figure 2.23(c), which shows that a major rise or geoid high exists east of the Louisville Ridge, between 38°S and 41°S. and 160°W and 150°W. It is elongated east–west, 1200km long by 300km wide, and is one of the larger rises of the western Pacific. In detail, it consists of three elevated regions, the western and central ones lying in an area which has been moderately well surveyed by echo-sounding. The known bathymetry of this surveyed area consists of smaller elevated features superimposed on a subtle regional high of no more than 500m. This appears to be inadequate to explain the large observed geoid anomaly, which is +1.0m to +1.5m over most of the area. In this case, either the bathymetric surveys are inadequate, or the crust must locally be anomalously dense.

Thus, satellite altimetry cannot give unequivocal information about ocean bathymetry. There must always be some control by more conventional shipborne techniques, to confirm or refute the interpretation.

Spreading patterns
The second way in which satellite altimetry helps the investigation of ocean floors relates to the main ocean ridges. Figure 2.23(a) is dominated by these features. The Central Indian and South-west Indian Ridges are both of slow-spreading type, while the South-east Indian Ridge is fast-spreading.

QUESTION 2.10 From your understanding of previous Sections, what is the correlation between spreading rate, topographic relief and geoid anomaly?

The picture is rather more complicated for the portion of the East Pacific Rise (EPR) shown on Figure 2.23(d). Between the Menard and Udintsev Fracture Zones the geoid high occurs up to 200km west of the bathymetric high. The charted ridge positions could be slightly in error but it is also possible that the geoid high *is* asymmetric relative to the ridge, possibly because of asymmetric spreading (different rates on each side of the ridge, Section 2.3.1), or even to a sideways shift of the spreading axis (ridge jumping, Section 4.2.2).

According to the simple correlations established above, where would you say the slowest-spreading regions on the East Pacific Rise and its continuation in the Pacific–Antarctic Ridge are to be found?

Near the Eltanin and Udintsev Fracture Zones and further west along the ridge, to the south of New Zealand. Clearly, there is some variation in spreading rate between different segments of this ridge system.

The greatest geoid anomaly on the Pacific–Antarctic Ridge occurs to the south of New Zealand, and this is where we would predict the slowest rate of sea-floor spreading. In fact, magnetic anomaly maps do demonstrate a decrease in spreading rate along the ridge, from about $10\,\text{cm}\,\text{yr}^{-1}$ north of the Menard Fracture Zone to $6\,\text{cm}\,\text{yr}^{-1}$ south of New Zealand.

Do you notice anything strange about the orientation of the traces of the Eltanin and Udintsev Fracture Zones on Figure 2.23(c) and (d) as you follow them south-eastwards away from the ridge?

Like most fracture zones, they show up as linear lows in the geoid surface, and they both undergo a sharp change in direction of about 20° at about the same distance from the ridge. This implies that a change must have taken place in the direction of sea-floor spreading. In fact, comparison with magnetic anomalies suggests that this occurred about 45–40Ma ago, and probably represents the same change in motion as that recorded by the bend in the Hawaiian–Emperor Chain (Section 2.5.3).

Figure 2.23 thus shows the distribution of some important bathymetric features and/or mass anomalies close beneath the sea-bed. These data provide a very useful basis upon which to plan shipborne investigations. Other Volumes in this Series describe how satellites are of immense use in other branches of oceanography.

2.7 SUMMARY OF CHAPTER 2

1   Aseismic (passive, Atlantic-type) continental margins develop where the continent and adjacent ocean basin form part of the same plate. They are underlain by thinned continental crust, upon which sediments accumulate to build up the continental shelf, slope and rise. Microcontinents are areas of continental crust split off from larger continental masses, which form either islands (crust of near-normal thickness) or submerged plateaux (thinned crust).

2   Seismic (active, Pacific-type) continental margins occur where a continent and the adjacent ocean basin belong to different plates, and oceanic lithosphere is being subducted beneath the continental margin. These are destructive plate margins. The continental shelf is typically narrower and the slope typically steeper than at aseismic margins, and a trench at the foot of the continental slope generally replaces the continental rise found at aseismic margins. Sediments accumulating in the trenches are partly scraped off to extend the inner trench wall oceanwards, and partly subducted. Island arcs occur at another kind of seismic (destructive) margin, being formed of volcanoes that lie above a subduction zone where one oceanic plate descends beneath another. They may be backed by small basins floored by oceanic crust that has formed by spreading somehow related to the nearby subduction.

3   Ocean ridges are the most important physiographic (bathymetric) features of the ocean basins. They are the constructive margins (spreading axes) of the plates, where new oceanic lithosphere is continually being generated. Gradients on the flanks of slow-spreading ridges are steeper than those on fast-spreading ridges. Median rift valleys are better developed (wider and deeper) on slow- than on fast-spreading ridges.

4   There is a systematic relationship between the depth to the top of the oceanic crust and its age of formation. This is the result of progressive cooling and subsidence of lithosphere with distance from the constructive margin. This relationship enables the age of oceanic crust to be estimated from its depth, and vice versa, and works to a good first approximation throughout the major ocean basins.

5   Transform faults offset spreading axes and lie parallel to small circles about the pole of relative rotation of lithospheric plates. They are seismically active, and they separate plates moving in opposite directions. Beyond the region of offset, they become fracture zones, and lie within a single plate, so seismic activity is much less. These faults and fractures result in scarps and clefts on the ocean floor. Major transform faults are also known as conservative plate margins, but some transform faults can have a component either of spreading (leaky transforms), or of subduction. However, the predominant sense of movement is always lateral.

6   Abyssal plains occupy large areas of the deep ocean floor. They are very flat due to burial of the rough topography of the oceanic crust by sediments. The sediments are either supplied by turbidity currents from adjacent aseismic margins (e.g. Atlantic Ocean), or mainly deposited from suspension in seawater (pelagic sediment), especially where trenches along seismic margins have prevented continent-derived material from reaching the plain (e.g. much of the Pacific Ocean).

7   Seamounts, oceanic islands and aseismic ridges are volcanic features rising from the ocean floor. Linear chains of seamounts and islands, as well as aseismic ridges, may result from hot-spot volcanism, whereby the oceanic plate moves over an intermittently or continually active localized source of magma rising from the deep mantle.

8   Satellite altimetry measurements can be used to map features on the sea-floor, because the mean sea-surface correlates with ocean bathymetry. Conventional bathymetric observations can be refined and extended using these data.

Now try the following questions to consolidate your understanding of this Chapter.

QUESTION 2.11   A seamount 1000m high has its summit 3000m below sea-level. What might its age be, and what assumptions do you need to make in arriving at its age? Would you expect the seamount to have a flat top?

QUESTION 2.12   Which of the following statements are true?

(a)   The age–depth relationship means that crust generated at a slow-spreading ridge must travel further before subsiding to a given depth than crust from a fast-spreading ridge.

(b)   Most island chains are linear hot-spot traces.

(c)   Sediment thicknesses increase away from ridge crests.

(d)   Continental crust below sea-level is thinner than normal.

(e)   Sediments in oceanic trenches become subducted.

QUESTION 2.13   In a particular ocean basin a spreading axis (ridge) runs north–south, but it is offset by an east–west transform fault which extends beyond the ridge offset as a fracture zone (Figure 2.25). To the south of the fracture zone, at a distance of 940km east of the ridge, the depth to the sea-floor is 5000m. Immediately north of here, on the other side of the fracture zone, the depth is 5500m. Use the age–depth relationship (Figure 2.13) to answer the following:

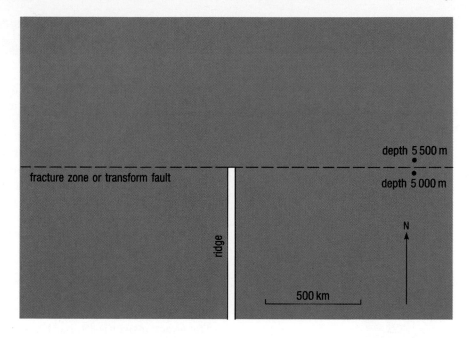

Figure 2.25 Incomplete map of the area discussed in Question 2.13. It will be useful for you to complete the map in answering Question 2.13(d).

(a) Which way is the ridge axis offset—to the left or to the right?

(b) What is the average spreading rate of this ridge?

(c) Is this therefore a 'fast' or a 'slow'-spreading ridge, and which of those ridges that you have encountered does it resemble?

(d) The position of the ridge north of the transform fault has not been surveyed. Assuming that spreading rate has remained constant, what is the amount of offset on the ridge due to the transform fault?

(e) How reliable do you think your figures are?

CHAPTER 3	THE EVOLUTION OF OCEAN BASINS

IMPORTANT: In this Chapter you will find frequent reference to the divisions of the geological time-scale. This is shown in the Appendix.

The Earth's oldest rocks—around 3800 Ma old—include both water-lain sediments and evidence of ancient oceanic crust. It follows that oceans have been forming since the beginning of the geological record, and probably before that. However, the shape of past ocean basins can only be worked out from observations of remnants preserved in continental areas. That is because ocean basins are relatively short-lived features of this planet: no oceanic crust older than about 160 Ma is known from the present oceans.

If we take the life cycle of a large ocean basin to average about 200 Ma, how many times could such basins have been formed since 3800 Ma ago?

The exact answer is 19, but because our figure of 200 Ma is only a crude guess and we do not know how rates of plate-tectonic processes in the distant geological past compared with those of the present, it is wiser to give an approximate figure of 15–20. This is probably a minimum, for the Earth's interior was a good deal hotter in the past than it is now, and the turnover of oceanic lithosphere could have been more rapid.

That simple calculation was designed only to give you a feeling for the time-scale of evolution of individual ocean basins, and it is obviously somewhat artificial. In the past, as one ocean basin expanded, another must have contracted, just as the Atlantic and Pacific are doing today. Thus, there is always some overlap in the history of different basins. Continents and ocean basins are continually changing their shapes and relative positions at rates that are geologically very rapid and are not slow even on human time-scales. The speed of sea-floor spreading has been compared with that of growing fingernails. Since the compilation of the first maps to cover any appreciable area of ocean, around five centuries ago, the Atlantic coasts have drawn apart from each other by about 10–20 m. That is a substantial movement, even though it represents only 0.0003% of the width of the ocean. Spreading rates in parts of the Pacific are several times greater than in the Atlantic.

3.1 THE EVOLUTION OF OCEAN BASINS

An individual ocean basin grows from an initial rift, reaches a maximum size, then shrinks and ultimately closes completely. Stages in this cycle are summarized in Table 3.1 and briefly reviewed below.

Whether or not the East African rift valleys really are an incipient ocean basin (Stage 1) and eastern Africa will eventually be split apart is debatable. Nevertheless, such rift valleys must develop along the line of continental separation. When separation does occur, sediments from the adjacent continents soon begin to build out into the new basin and will become part of the eventual continental shelf–slope–rise zone. As the

Table 3.1   Stages in the evolution of ocean basins, with examples.

Stage	Examples	Dominant motions	Characteristic features
1 embryonic	East African rift valleys	crustal extension and uplift	rift valleys
2 young	Red Sea, Gulf of California	subsidence and spreading	narrow seas with parallel coasts and a central depression
3 mature	Atlantic Ocean	spreading	ocean basin with active mid-ocean ridge
4 declining	Pacific Ocean	spreading and shrinking	ocean basin with active spreading axes; also numerous island arcs and adjacent trenches around margins
5 terminal	Mediterranean Sea	shrinking and uplift	young mountains
6 relict scar	Indus suture in the Himalayas	shrinking and uplift	young mountains

spreading axis builds away from the marginal areas, the sediment supply dwindles (Stage 2). The continents become increasingly distant from the hot and relatively elevated spreading ridge. The intervening regions subside by thermal contraction of the underlying lithosphere (Figure 2.13), abyssal plains form, and the continental shelf–slope–rise zone becomes fully developed. The continental margins are sub-parallel to the central spreading ridge, as in the Atlantic (Stage 3).

Stage 4 involves the development of one or more destructive plate margins. The reason for the formation of new destructive margins probably lies in changing circumstances on another part of the globe, such as continental collision or the initiation of new continental rifting. If (as is almost certain) the Earth is neither expanding nor contracting, the net rates of spreading and subduction over any great circle on the Earth must be equal, and the pattern of plates and plate motion must adjust to keep this so.

The Mediterranean is an ocean in the final stages of its life (Stage 5), with the African plate being consumed under the European plate. Unless the world system of plates changes so that the northward movement of Africa relative to Europe is halted, the continental blocks of Europe and Africa will eventually collide, and new mountain ranges will form (Stage 6).

QUESTION 3.1   Study Figure 3.1.

(a)   Tethys was a major ocean basin, a branch of Panthalassa, that once separated Eurasia from the southern continent, Gondwanaland. How have its shape and size changed over the past 170 Ma and what are its present-day marine remnants?

(b)   How long ago did the Atlantic Ocean start to open?

(c)   When did the Indian Ocean start to open?

(d)   Was the period from 170 to 100 Ma or the one from 100 to 50 Ma more significant in terms of the fragmentation of Gondwanaland?

(e)   What were the major changes between 50 Ma and the present day?

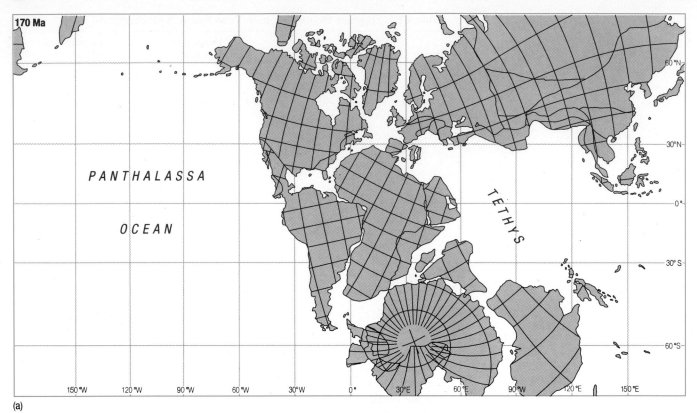

(a)

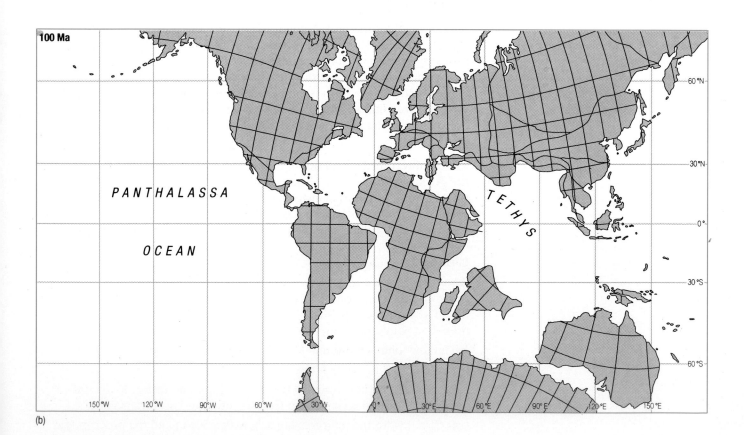

(b)

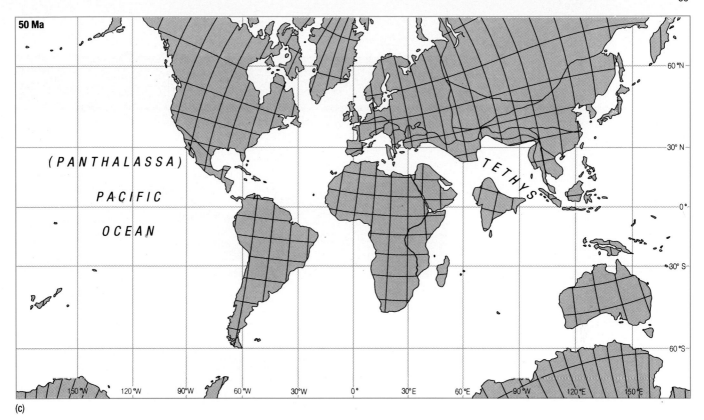

50 Ma

(PANTHALASSA)

PACIFIC

OCEAN

TETHYS

60 °N
30° N
0°
30° S
60 °S

150 °W   120 °W   90°W   60 °W   30°W   0°   30°E   60 °E   90 °W   120 °E   150 °E

(c)

Figure 3.1   Palaeogeographic reconstruction, compiled from topographic, palaeoclimatic and palaeomagnetic data. Panthalassa was the huge ocean that dominated one hemisphere. Pangaea was the supercontinent in the other hemisphere, of which Eurasia and Gondwanaland were two components. (a) Jurassic, about 170Ma ago. (b) Cretaceous, about 100Ma ago. (c) Eocene, about 50Ma ago. The maps show *present-day* coastlines for ease of reference. Ancient coastlines did not coincide with these.

3.2   THE BIRTH OF AN OCEAN

Figure 3.2 summarizes the development of a new ocean basin. During crustal extension, the plastic lower part of the crust is stretched, but the brittle upper part is rifted. Blocks of crust slide down fault planes, and sediments accumulate in lakes and valleys which occupy the resulting depressions. When separation occurs, basaltic magma rises to fill the gap between the two continental blocks. As the resulting new oceanic crust is both thinner and denser than continental crust, it lies below sea-level. The remainder of the lithosphere, below the crust, is composed of upper mantle material.

Initially, the young marine basin is fairly shallow. If repeated influxes of seawater become wholly or partly evaporated, salt deposits (**evaporites**) will accumulate on the new sea-floor. Otherwise, there will be normal marine sedimentation of muds, sands and limestones, depending on local conditions. One of the clearest examples of a young ocean basin is the Red Sea.

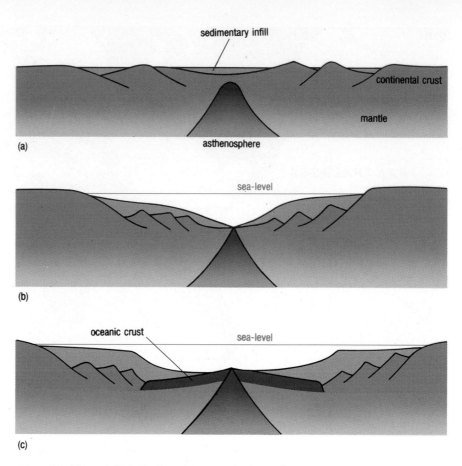

Figure 3.2 Diagrams illustrating how a new ocean basin may form.

(a) The surface of the stretched and rifted region is still above sea-level and may even be uplifted due to thermal expansion because of a heat source at the site of the future spreading axis. Terrestrial sediments occupy the rift valley.

(b) The future continental margins have thinned enough to subside below sea-level and marine sedimentation has begun. Sediments thin towards the new spreading axis.

(c) Separation is complete, a new spreading ridge has developed, and a shelf–slope–rise zone is forming (*cf.* Figure 2.6).

3.2.1 THE RED SEA

There are two main physiographic provinces in the Red Sea: a narrow deep axial zone and broad shallow areas which flank it on either side (Figure 3.3(a)). Miocene evaporites, deposited between about 20 and 5 Ma ago and over 4 km thick in places, underlie the shallower waters of the flanking regions. They obscure the nature of the crust beneath, which may be either oceanic or continental (Figure 3.3(b)).

The evaporites were deposited at a time when the only marine connection from the Red Sea was to the Mediterranean by an intermittent, shallow, seaway. Evaporite deposition ended when this seaway was finally broken and a new connection with the Indian Ocean was opened up in the south, about 5 Ma ago at the end of the Miocene. Open water conditions were established, in which planktonic organisms flourished, especially in the southern Red Sea. High rates of biogenic (organically derived)

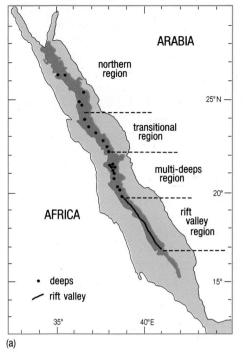

(a)

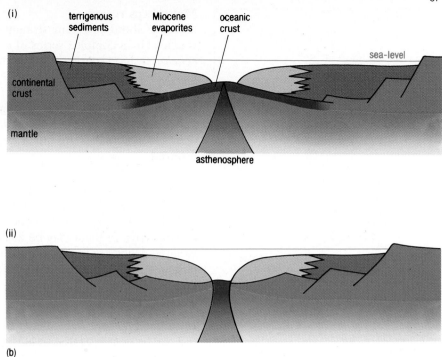

(i)

(ii)

(b)

Figure 3.3(a) Outline map of the Red Sea, with the axial zone (dark blue) defined by the 500-fathom isobath, and subdivided into four main sections, described in the text. (1 fathom = 6 feet = 1.83m.)

(b) Highly schematic cross-sections to illustrate two possible configurations for the crust beneath the sediments of the flanking regions in the Red Sea.

sedimentation caused bathymetric features to be smothered, and they become much less obvious south of about 16°N.

Further north, the post-Miocene biogenic sediments give way to a thinner sequence of terrigenous (land-derived) clays, sands and gravels. These were derived by erosion of the flanks of the basin, and can also be found interbedded with the Miocene evaporites, especially near the margins.

Only in the axial zone, which represents that part of the Red Sea generated since the end of evaporite deposition, can we observe the immediate effects of sea-floor spreading. On the basis of seismic and magnetic surveys, submersible observations and side-scan sonar mapping, the axial zone can be subdivided into several regions along its length (Figure 3.3(a)), as described below.

Rift valley region
The southern part of the axial trough is now known to have a well-developed straight central rift (similar to that on the Mid-Atlantic Ridge; Figure 2.12), which is offset by 3–10km about every 30–50km. These offsets may be either transform faults or some sort of non-transform offset such as overlapping spreading centres (Section 4.2).

High-amplitude linear magnetic anomalies occur throughout this region, though they become weak and irregular at the offsets. Measurements of the magnetic anomaly stripes indicate that spreading has proceeded at a rate of about 0.8cmyr^{-1} for about the past 5Ma. In some places it may have been asymmetric, spreading faster to the east than the west.

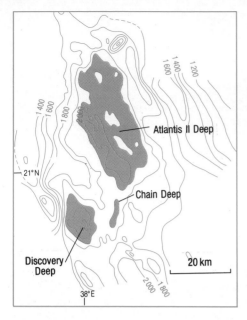

Figure 3.4 Bathymetric details of some major deeps in the multi-deeps region of Figure 3.3(a). Hot, metal-rich brines are found in them, and metalliferous muds are being deposited there. Depths are given in metres.

Multi-deeps region

North of about 20°N, the straight axial rift loses its identity and is replaced by a complex series of axial deeps, distributed partly in an *en echelon* fashion, perhaps due to offsets by transform faults. The deeps are best developed between about 20°N and 22°N and they have attracted commercial interest on account of the metal-rich hot brines and muds which the active ones contain (Figure 3.4). Individual deeps have a rift-valley type structure with strong magnetic anomalies, but between the deeps the anomalies are much weaker and the axial region is sediment-covered.

Transitional and northern regions

Beyond about 22°N, the deeps become progressively narrower and less well developed, and the associated magnetic anomalies suggest that the oceanic crust in them may be 2Ma old or less. North of about 25°N only isolated deeps are found, the high-amplitude magnetic anomalies characteristic of oceanic crust which occur further south have virtually disappeared, and the region appears to have a more or less continuous sediment cover.

In summary, then, only about 80km width of new ocean floor up to 5Ma old can be demonstrated to have formed in the southern part of the axial zone; further north, ocean floor has formed only in the deeps and is 2Ma old or less.

All of this suggests strongly that the axial zone of the Red Sea is a northward-propagating zone of separation between adjacent lithospheric plates. The fracture began to open properly about 5Ma ago in the south but has yet to do so in the north. This is consistent with the end of evaporite deposition in the Red Sea about 5Ma ago, when a link with the Indian Ocean was established via the Gulf of Aden.

That is all very well, but what does it tell us about the nature of the crust outside the axial region of the Red Sea?

It tells us rather little. Figure 3.5 shows a typical magnetic anomaly profile for the southern Red Sea. The well-defined axial anomalies give way to an irregular and much weaker pattern in the flanking regions, where the thick sediments have 'muffled' or attenuated the signals. It is not possible to decide whether the flanking anomaly pattern is due to basaltic oceanic crust (Figure 3.3(b)(i)) or to stretched and thinned continental crust (Figure 3.3(b)(ii)), injected by basaltic dykes.

Even the history of the Red Sea area as a whole (including the Gulf of Aden), as inferred from both geological and geophysical investigations over several decades, still does not permit an unequivocal choice between these alternatives. It seems likely that the first stage, about 40Ma ago, was the propagation of a crack from the Arabian Sea westward through the Gulf of Aden and northward through the Red Sea towards the Gulf of Suez, associated with development of a rift system within the continental crust. About 25Ma ago a new fracture developed in the north along what is now the Gulf of Aqaba/Dead Sea line, running towards the north–north-east. Transcurrent movement along this line accompanied further widening of both the Gulf of Aden and the Red Sea as Arabia moved away from Africa. Oceanic crust formed in the Gulf of Aden.

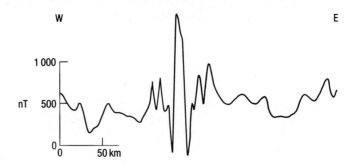

Figure 3.5   Magnetic anomalies for the southern Red Sea, showing the contrast between the strong central pattern over the axial trough and the subdued pattern on either side. For this profile, correlation with the magnetic reversal time-scale (black = normal polarity; white = reverse polarity) gives an average full spreading rate of $1.5\,cm\,yr^{-1}$ over the past 4.5 Ma. The magnetic field is measured in nT (nT = nanotesla = 10^{-9} T).

Along the Red Sea rift, however, either the continental crust continued to stretch and subside, or it separated and oceanic crust formed. About 5 Ma ago the system was reactivated when transcurrent movement was renewed along the Gulf of Aqaba/Dead Sea line, sea-floor spreading continued in the Gulf of Aden, and in the Red Sea new oceanic floor was generated either for the first time or as a continuation of an earlier phase of sea-floor spreading.

If the alternative in Figure 3.3(b)(i) proves to be correct, then the Red Sea opened and spread for a period of time from about 25 Ma ago and then stopped spreading. Deposition of Miocene evaporites occurred during and after this phase of spreading. Renewed spreading began about 5 Ma ago in the south and has propagated northwards since.

If the alternative in Figure 3.3(b)(ii) is correct, then crustal thinning and subsidence occurred before and during the Miocene, and the evaporites were deposited on thinned continental crust. Final separation occurred about 5 Ma ago in the south and has propagated northwards since.

The alternative models for the Red Sea will not be resolved until boreholes are put down through the sediments of the flanking regions. Meanwhile, we move on to consider larger ocean basins now in Stage 3 or 4 of Table 3.1, bearing in mind that in the early stages of formation, they must have resembled the Red Sea.

QUESTION 3.2   In which direction would you expect to find the pole of relative rotation about which the African and Arabian plates are moving to form the Red Sea?

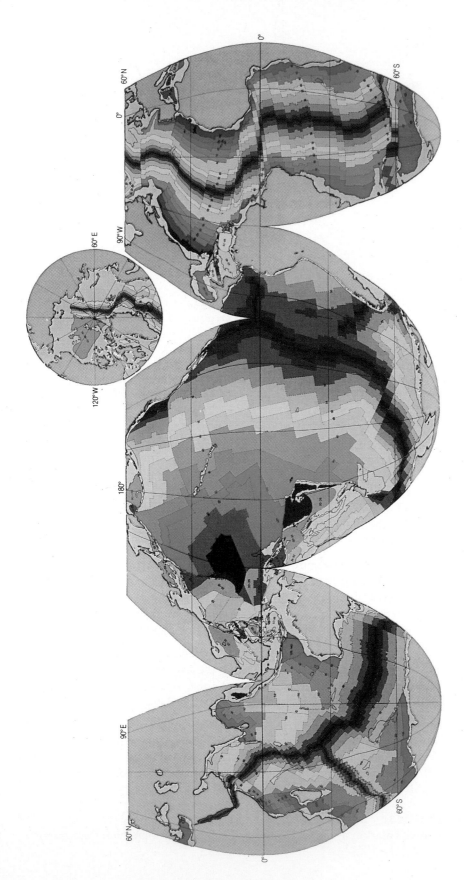

Figure 3.6   The age of the ocean floor, showing strips of floor of different ages derived mainly from measurements of magnetic anomaly stripes. Boundaries are drawn at 2, 4, 9, 20, 35, 52, 65, 80, 95, 110, 120, 140 and 160 Ma intervals in a colour scheme that runs from red (youngest) through yellow and green to blue (oldest).

3.3 THE MAJOR OCEAN BASINS

Figure 3.6 summarizes the age distribution of oceanic crust beneath the world's oceans, as determined from magnetic anomaly patterns. The virtually symmetrical pattern of ages about the ocean ridges is visible everywhere, and it is clear that (apart from the Caribbean area and the extreme south-west) the Atlantic has had the least-complicated evolution of any of the three main ocean basins. The Pacific and Indian Oceans display a more complex history partly because of the development of major subduction zones along one or more boundaries and partly because of adjustments in spreading direction.

From Figure 3.6, it is relatively easy to reconstruct stages in disruption of the continental jigsaw puzzle for the Atlantic. It is simply a matter of moving the continents back along the transform faults to determine the positions of their margins at any particular time, as represented by the age of the ocean floor magnetic stripes.

The Pacific and Indian Oceans are more difficult to close up. The Pacific is almost surrounded by subduction zones, so much of the evidence of its older history has disappeared. The East Pacific Rise runs into the North American continent, beneath which its northern end has been subducted. The widths of the ocean floor strips increase northwards along the line of the East Pacific Rise in a way that is consistent with increasing spreading rates from south to north. You have already seen evidence of changes in spreading direction in the north-western Pacific (Figure 2.21), and more evidence is to be seen in the pattern of ocean floor ages in Figure 3.6.

What does the fact that the Hawaiian Chain is oblique to the spreading trend indicated for 0–43 Ma Pacific ocean floor by the age strips in Figure 3.6 tell you about the movement of the East Pacific Rise? Bear in mind that hot spots can be regarded as effectively stationary with respect to the Earth as a whole.

The East Pacific Rise must have been moving relative to the frame of reference provided by the Hawaiian and similar chains (the 'hot spot reference frame'). It is important for you to realize that magnetic anomalies and the sea-floor age strips which can be mapped using them record only the motion relative to the spreading axis at which the sea-floor was formed, and that it is quite possible for the spreading axis itself to be migrating relative to the deep Earth. In fact, it is impossible to have several spreading axes on a sphere and keep them all stationary, unless the plate motion at each one is individually and exactly compensated for by a nearby destructive plate margin. Thus, a spreading axis can migrate across an ocean basin, and will be destroyed if it gets carried into a subduction zone. Figure 3.7 shows how the global plate boundaries are thought to have evolved over the past 61 Ma based on the types of data you have read about in this Chapter and the previous one, especially hot-spot traces and magnetic anomaly patterns.

The oldest oceanic crust in the Pacific is found in the north-west, but the western Pacific as a whole is an area of great complexity. This is because of the generation of new oceanic lithosphere at various spreading axes above subduction zones, where island arcs are being built and then split

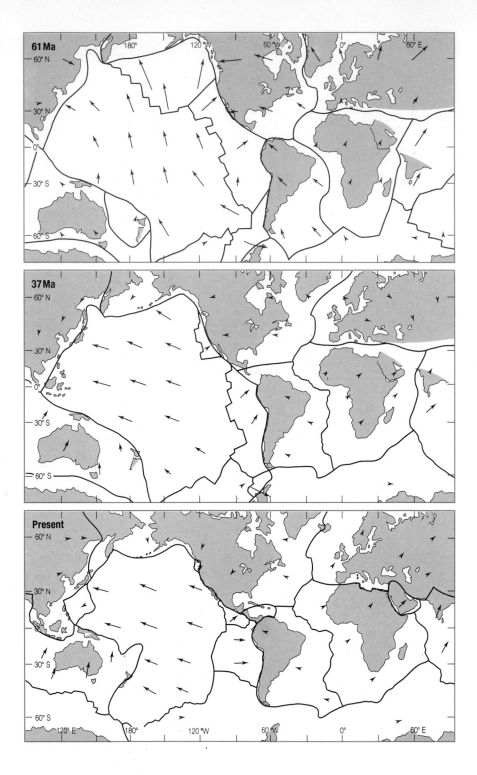

Figure 3.7   Evolution of plate boundaries over the past 61 Ma. The arrows show the directions and relative rates of motion of the plates.

apart and back arc basins are forming, as outlined in Section 2.2.2 and illustrated in Figure 3.8. These events occur independently of sea-floor generation at the East Pacific Rise.

In the Pacific, there is an added complication when it comes to reconstruction of the continental jigsaw puzzle. Palaeomagnetic, geological and palaeontological evidence suggest strongly that substantial tracts of western North America are 'suspect terranes'. These are blocks

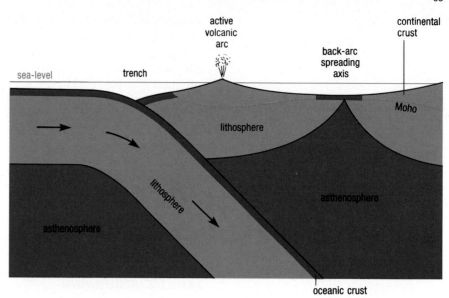

Figure 3.8 A schematic cross-section through a typical western Pacific continental margin. A small spreading centre has generated new oceanic crust behind the island arc. In some places, an even more complex series of island arcs and ocean basins separates the deep ocean trench from the continental margin.

of continental crust—microcontinents—that have been transported by plate movements for great distances across the Pacific to become accreted onto North America.

The Indian Ocean has several features of interest, too. Its northern boundary is a major complex subduction zone, represented by the Himalayan belt and the Java trench system. South of India, the spreading direction changed from north–south to north-east–south-west about 50 Ma ago when the present South-east Indian Ridge became established.

The aseismic Ninety-east Ridge (Section 2.5.4) must lie on the line of an old major transform fault, for the age of the crust changes in opposite directions on either side of it.

QUESTION 3.3 (a) In what directions do crustal ages change on either side of the Ninety-east Ridge?

(b) Were these areas of oceanic crust generated before or after development of the present South-east Indian Ridge?

(c) What happened to the spreading axis which generated the crust immediately east of the Ninety-east Ridge?

The South-west Indian Ridge also has a complex history. It was originally a major transform fault transecting the Carlsberg/South-east Indian Ridge, but about 20 Ma ago it became a spreading ridge.

How can we tell that?

Age boundaries of ocean floor older than about 20 Ma trend at right angles to the South-west Indian Ridge and are offset across the line of the ridge, but they are parallel to the ridge segments after about 20 Ma.

You can find other examples of changes of spreading rate and direction, and development of new spreading axes and subduction zones, displayed in Figure 3.6. Look, for example, to the west of Australia. You may also have noticed that many of the ocean floor age strips are oblique to subduction zones, e.g. in the north-east Indian Ocean and parts of the western Pacific. Oblique subduction of ocean floor is by no means exceptional, and it means that when continent–continent or continent–island arc collision occurs, it will not necessarily be head-on, and major transcurrent faulting may result.

In the major ocean basins, irrespective of whether they are classified as Stage 3 or 4 in Table 3.1, there is no indication that increasing age is correlated with any decline in the intensity of sea-floor spreading activity. The Pacific basin is the oldest, for instance, but it has the fastest spreading rates. When we come to Stage 5, we find that even in the latest stages of evolution, there is little diminution of vigour.

3.3.1 THE MEDITERRANEAN

The Mediterranean can be classified as an ocean in the final stages of its life cycle, the only major remnant of the once-extensive Tethys Ocean (Table 3.1, Stage 5; Figure 3.1). The Mediterranean is shrinking as the African plate continues to thrust its way northwards beneath the European plate. We might therefore expect that the Mediterranean would be floored by oceanic crust dating back perhaps as far as Jurassic times, which would be consistent with its being the remnant of an old ocean, and that it would have an obvious major trench. It turns out that the Mediterranean has been broken into many minor plates, whose boundaries must be delineated partly by analysis of earthquakes, which are sporadic and scattered in this region; and partly by drilling. The deep basins contain several kilometres of sediments, including evaporites (see Chapter 6), and this hampers geophysical investigations into the nature of the underlying crustal layers.

The eastern Mediterranean is floored by crust of Cretaceous age (c. 110Ma), bordered by an area of middle Tertiary (25Ma) crust south of Italy, possibly the result of back-arc spreading. There is a collision zone running south of Cyprus, where the African plate meets the Eurasian plate, and there is some evidence of subduction there. Miocene and younger (10–2Ma) oceanic crust in the western Mediterranean is generally believed to have formed in a back-arc setting, associated with the subduction that removed older ocean floor from this region. Active volcanism and frequent earthquakes in and around the Mediterranean show that it is still evolving.

QUESTION 3.4 From your understanding of Section 2.6, can you see any correlation between crustal age and shape of the geoid in the Mediterranean (Figure 1.16)?

Apart from the eastern Mediterranean sea-floor, ocean floor older than about 70Ma is represented in the Mediterranean region also by slivers of oceanic crust and upper mantle which have been tectonically removed from the ocean floor and emplaced over a continental margin, usually during a collision event. A piece of ocean floor which has been preserved in this way is called an **ophiolite** or **ophiolite complex**. Ophiolites are too small and fragmentary to preserve magnetic stripes, but they can be dated by other methods (e.g. radiometrically or by examining the fossils in the oldest sediments associated with them). Ophiolites of mostly late Cretaceous age (*c.* 80 Ma) are preserved in collisional mountain chains to the north of the present Mediterranean in the Balkans and Asia Minor. This belt runs eastwards all the way through the Himalayas and is the remains of the Tethys Ocean. Older ophiolites tend to occur near the western Mediterranean, particularly in the Alps, and range back to the late Triassic (*c.* 220 Ma). These may represent the oldest ocean floor of the Mediterranean–Tethys region.

Reconstruction of the evolution of ocean geometry is more than just an academic exercise. Combined with information from the sediments on the ocean floor, it can be used to reconstruct past climatic and circulation patterns. Compilations such as Figure 3.6 are continually being up-dated as less well-known areas are surveyed in greater detail, sometimes with the help of satellite bathymetry data. The more we understand about fluctuations in past oceanic cycles, the better we shall understand the present-day oceans. We shall look at this with particular reference to sea-level in Chapter 6, but next we turn to the mechanisms of generation of new oceanic crust.

3.4 SUMMARY OF CHAPTER 3

1   Oceanic crust is much younger than most continental crust. Oceanic lithosphere must have been generated (at ridges) and destroyed (at subduction zones) many times since the formation of the Earth. Ocean basins form by stretching and splitting (rifting) of continental crust, and the rise of mantle material into the crack to form new oceanic lithosphere.

2   The Red Sea is an embryonic ocean that appears to be opening progressively from the south, where the axial region is underlain by oceanic crust and has a rift valley. Further north are isolated deeps—with metal-rich muds—and there is less evidence of oceanic crust in the axial region. It is not known for certain whether the thick evaporites bordering the axial region rest on oceanic crust or thinned continental crust.

3   Among the major ocean basins, the Atlantic has the simplest pattern of ocean-floor ages, which shows that it has opened fairly steadily since its birth. Subduction is confined to relatively small island arc systems in the Caribbean and the extreme south-west. Successive stages in the shape of the Atlantic basin are therefore fairly easy to reconstruct, by moving the continents back at 90° to the magnetic anomaly stripes.

4   In contrast, both the Pacific and Indian Oceans (which have major subduction zones) are characterized by changes of spreading rate and direction and the development of new spreading axes. Because of these complications, it is difficult to reconstruct how the shapes of these ocean basins have changed with time. The occurrence in western North America of 'suspect terranes', which in some cases are believed to have originated as microcontinents in the south-west Pacific, make attempts at such reconstruction even more complicated.

5   The Mediterranean represents an ocean in the final stages of its life cycle, contracting as Africa pushes northwards into Europe and western Asia. However, no oceanic crust older than the Cretaceous is known from the Mediterranean basin, and crust as young as 2Ma has been found there, demonstrating that there is no inverse correlation between age of an ocean basin and the vigour of sea-floor spreading and plate-tectonic activity.

Now try the following questions to consolidate your understanding of this Chapter.

QUESTION 3.5   'India moved northwards at rates of between 10 and 20cmyr^{-1} from about 135 to about 45Ma ago. For the next 25Ma or so, it moved more slowly at about 5cmyr^{-1}.' Can these statements be *qualitatively* correlated with information in Figure 3.6?

QUESTION 3.6   What is the approximate ratio between the average spreading rates over the past 52Ma for the East Pacific Rise at the Equator relative to the Mid-Atlantic Ridge at 30°S? If the average spreading rate in the Atlantic was 2cmyr^{-1} over this period, what was it for the East Pacific Rise?

CHAPTER 4
THE STRUCTURE AND FORMATION OF OCEANIC LITHOSPHERE

Knowledge of the nature of the oceanic crust and upper mantle comes from four main sources: (1) geophysical techniques, notably seismic refraction and reflection, as well as magnetic and gravity surveys and heat flow measurements; (2) examination and measurement of physical properties of rocks dredged from the sea-bed and cored from the upper parts of the crust; (3) direct observation and photography of the sea-bed using submersibles; and (4) land-based studies of ophiolite complexes. It was seismic refraction studies which first revealed that the oceanic lithosphere consists of layers whose seismic velocity (the speed at which sound waves travel through the rock) increases with depth. These seismic layers are numbered from 1 (at the top) to 4, and have been matched up with rock types according to the scheme shown in Figure 4.1.

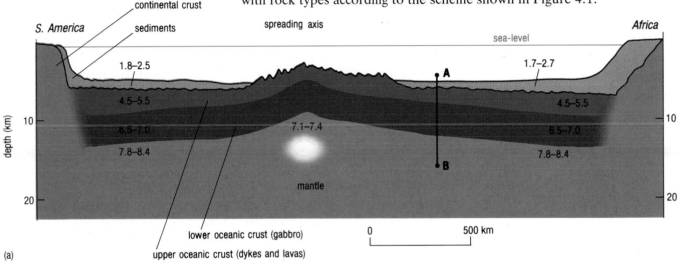

(a)

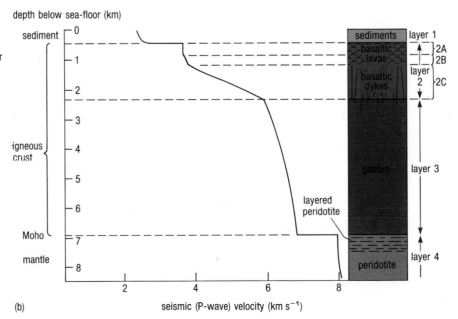

(b)

Figure 4.1(a) Highly schematic sketch cross-section through the Atlantic Ocean with great vertical exaggeration. Velocities are given in km s^{-1}. See (b) for an explanation of the velocity layers.

(b) Typical seismic structure of the oceanic crust and upper mantle, showing how the velocity of propagation of P-waves increases with depth. The column is from the vertical line A–B in (a). P-waves are compressional waves, analogous to sound waves in air. The seismic velocity layers are usually matched with rock type as on the right; see text for discussion. The individual rock types named in the column in Figure 4.1(a) are illustrated in Figure 4.2.

The following points summarize the layered structure of the ocean floor:

(i)   The igneous layers (2–4) are formed by processes at spreading axes, and are progressively buried by the sediments of layer 1, which eventually become consolidated.

(ii)   Layers 2 and 3 constitute the oceanic crust. The normal total thickness of these layers is 6–7 km everywhere (except beneath active spreading axes and near transform faults and fracture zones).

(iii)   The two most important changes in seismic velocity are at the base of the layer 1 sediments and at the base of layer 3. The latter is the Mohorovičić discontinuity (the **Moho**), where gabbro at the base of the crust rests on peridotite at the top of the mantle.

(iv)   Within layers 2 and 3 there are no definite seismic discontinuities, but rather a general increase of seismic velocity with depth. Subdivisions within layer 2 and the distinction between layers 2 and 3 are defined by changes of velocity gradient (changes in the rate of increase of seismic velocity with depth), and not by abrupt changes of seismic velocity.

(v)   Despite their different appearances (Figure 4.2), the rocks of layers 2 and 3 are virtually identical in overall chemical and mineralogical composition. The minerals of which they are composed are plagioclase feldspar, pyroxene and olivine in order of decreasing abundance. Layer 4 is different in both chemical and mineralogical composition, which accounts for the seismic discontinuity at the base of layer 3 (the Moho). The principal minerals of the peridotite of layer 4 are olivine and pyroxene.

(vi)   The subdivisions of layer 2 shown in Figure 4.1 cannot always be recognized, but where they are, drilled samples suggest the following explanation. In sub-layer 2A the pillow lavas (see Section 4.1.1) are mostly broken and rubbly, with many fractures and voids. In sub-layer 2B, the pillows are generally less broken and the voids are mostly filled with clays and other alteration minerals. This is because as soon as oceanic crust is formed it becomes subject to chemical alteration and metamorphism, through interaction with seawater. The higher the temperature, the more intense these interactions become. Much of the oceanic crust has in fact undergone some degree of alteration and metamorphism since its formation, and some would argue that the distinctions between layers 2B and 2C, and between layers 2C and 3, should be matched with differences in the intensity of alteration and metamorphism rather than differences in rock type. We shall consider sea-floor alteration and metamorphism and its causes in Chapter 5.

(vii)   Seismic velocities increase more rapidly with depth in layer 2 than in layer 3, chiefly because the pressure of successive layers of lava tends to close up cracks and fissures in the underlying rocks. Seismic velocities in layer 2 also tend to increase with age, i.e. with distance from the ridge, as the remaining cracks and fissures and pore spaces become infilled with minerals formed as a result of the interactions between rock and seawater.

QUESTION 4.1   Why should fractured lavas have lower seismic velocities than equivalent unfractured rocks?

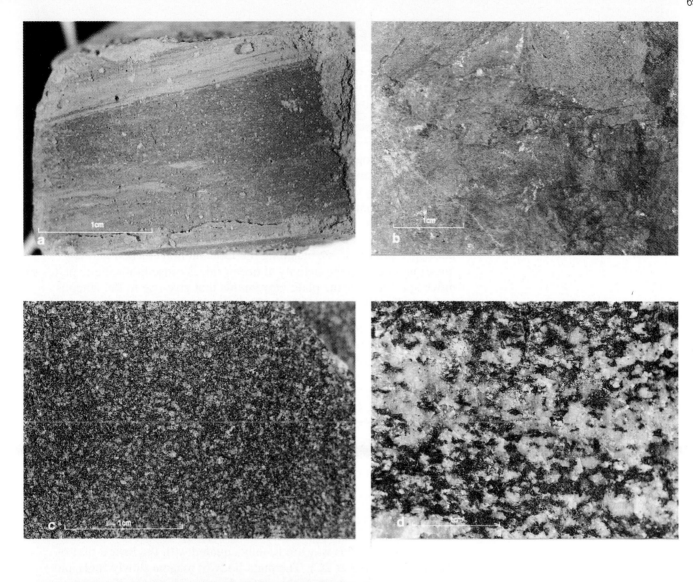

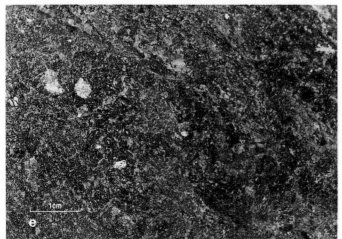

Figure 4.2   Colour photographs of sediments and igneous rocks of the oceanic crust and mantle.

(a)   Deep-sea muds (oozes) (layer 1).

(b)   Fragment of basaltic pillow from layer 2.

(c)   Fragment of basaltic dyke from layer 2.

(d)   Gabbro from layer 3.

(e)   Peridotite from layer 4.

4.1 THE FORMATION OF OCEANIC LITHOSPHERE

The axes of the ocean ridge systems are the most active volcanic zones on Earth, where new oceanic lithosphere is generated at rates of between 10 and 200 km per million years ($1–20 \, cm \, yr^{-1}$).

Igneous processes at ridge crests are remarkable in two other important respects. Unlike volcanoes elsewhere, which are active for relatively short periods of geological time, the volcanic activity at ocean ridges is episodically continuous for much of the lifetime of an ocean basin, which may last as long as a few hundred Ma. Moreover, the generation and eruption of magma occurs usually in such a consistent fashion that the same sort of layering can be traced over hundreds of millions of square kilometres of ocean floor. Much still remains uncertain about the sea-floor spreading process, including the fundamental chicken-and-egg question: does igneous activity at ocean ridges cause lithospheric plates to move apart, or is it the plate movements that give rise to the igneous activity? The answer is probably the latter.

Whatever the uncertainties about the details, there is broad agreement on the overall mechanism (Figure 4.3). The mantle beneath the lithosphere—the asthenosphere—behaves plastically because it is very near to its melting point, and rises upwards beneath ocean ridges to fill the gap between diverging lithospheric plates. As the asthenosphere rises beneath the ridge crests, the peridotite, of which it is composed, partially melts to form basaltic magma that accumulates at the base of the crust in large **magma chambers**, whence it is erupted along axial fissures to form pillow lavas on the ocean floor. Eruption at any one place along the axis is episodic, and at the end of each local episode some magma solidifies inside the fissure to form a vertical sheet of rock (a **dyke**), typically about a metre in width. The next eruptive event in the same area occurs by the formation of a new fissure near the previous one, sometimes by the splitting apart of the dyke which filled the previous fissure. The 'sheeted dykes' formed in this way are usually equated with the lowest part of seismic layer 2 (layer 2C). The main body of magma slowly cools and crystallizes at the edges, as the plates diverge. This crystallized magma forms the gabbro of layer 3 and the layered peridotite in the top of layer 4. The centre of the magma chamber over the axis probably remains liquid for much of the time due to replenishment by fresh magma from below. The unmelted residue of the mantle accretes to the diverging plates deeper down, to form the rest of the oceanic lithosphere.

QUESTION 4.2   Assuming that dykes which are formed during spreading are on average 1 m wide, about how often on average is a dyke formed at a given part of the Mid-Atlantic Ridge (half-spreading rate $2 \, cm \, yr^{-1}$) and the East Pacific Rise (half-spreading rate $8 \, cm \, yr^{-1}$)?

4.1.1   PILLOW LAVAS: THE TOP OF THE OCEANIC CRUST

Pillow lavas usually develop wherever basaltic magma erupts under water, so the top of the oceanic crust is nearly everywhere made of pillow lavas—the most abundant volcanic rock on Earth.

Would you expect lavas that erupt under water to cool more or less rapidly than those that erupt in air?

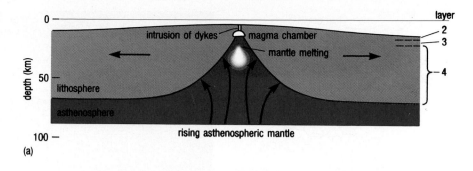

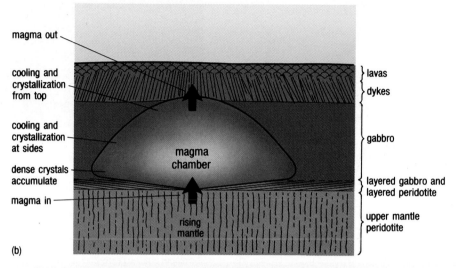

Figure 4.3(a) A cross-section showing igneous processes at active ridge crests. Rising asthenosphere fills the gap between separating plates. Some of it melts, ultimately giving rise to lava eruptions at the sea-floor and the formation of new oceanic crust. The remainder accretes to the edges of the lithosphere.

(b) Diagrammatic enlargement of the axial region of (a). Note that much of the magma crystallizes in the magma chamber itself. You are not expected to understand all the details of this diagram.

Lavas that erupt under water cool more rapidly, because water has greater thermal conductivity and specific heat than air. This rapid cooling is perhaps the chief reason why underwater lava forms into distinctive pillow-like shapes: as soon as the rising magma reaches the sea-floor, a chilled rind forms on the outer surface, which is still flexible enough to change shape as the lava flows. As more magma arrives at the surface to feed the flow, pressure inside the rind causes bulbous or tubular protrusions to form, which commonly become detached from the main flow when they reach a critical size, and form separate pillows. The formation of pillow lavas has been the subject of much detailed study. Figure 4.4 (overleaf) summarizes the main stages in the growth of pillows. By repeating this process, piles of pillows, up to hundreds of metres thick, can develop, to produce a characteristic hummocky topography on the sea-floor. Typical pillows range from about 10cm to over a metre in diameter, and up to several metres in length. When solidified, they have dark glassy rims (the original chilled rind) enclosing a more crystalline interior which cooled more slowly and in which radially arranged cooling cracks (joints) are often well developed.

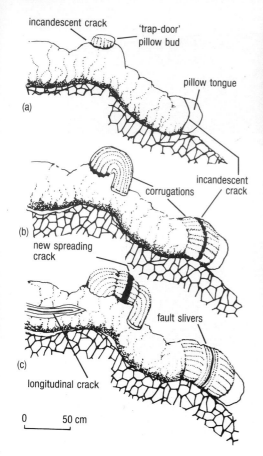

incandescent crack   'trap-door' pillow bud

(a)

pillow tongue

corrugations   incandescent crack

(b)

new spreading crack

fault slivers

(c)

longitudinal crack

0   50 cm

(d)

Figure 4.4(a)–(c)   Typical stages in the development of lava pillows, showing how they enlarge by the growth of new outer crust adjacent to incandescent cracks.

(a)   A pillow tongue growing down the slope is connected to a larger feeder-flow coming in from the left. Incandescent cracks in the outer thin crust form in response to tension induced in the crust by pressure of fluid lava within the tongue, and a small 'trap-door' pillow bud grows out from a circular crack at a weak point.

(b)   New lava with a corrugated crust is extruded from the crack. The pillow tongue lengthens down the slope, and the pillow bud continues to grow, eventually bending down under the influence of gravity (*cf.* part (c), which shows the corrugated surface of a curved pillow).

(c)   The spreading along the crack at the end of the pillow tongue slows down, causing a shift from the production of corrugations to the production of fault-slivers parallel to the spreading crack. A slowly widening longitudinal crack opens on the thick part of the pillow tongue. A new spreading crack develops in the middle of the bud, causing it to lengthen still further.

(d)   A submarine photograph taken during the FAMOUS project (Section 4.1.2) showing the sampling arm of the submersible (Figure 1.21) collecting fragments of lava forming the ocean floor.

Results from boreholes show that the volcanic layer does not consist entirely of pillow lavas. Massive (unpillowed) sheet flows also occur, sometimes interlayered with pillowed flows. In addition, pillows can easily fragment, so there are often local accumulations of volcanic breccia (debris) made of broken pillows and fragments of chilled glassy rind.

The formation of pillow lavas thus makes for an irregular sea-floor and tectonic and other processes combine to make the topography even more rugged (*cf.* Figures 2.11 and 2.12).

4.1.2   FORMATION OF THE VOLCANIC LAYER—A CASE STUDY

This Section presents the results of a highly detailed and now classic survey of a part of the Mid-Atlantic Ridge, made during the 1970s. It provided a major breakthrough in understanding the pattern of volcanism, fissuring, rifting (faulting), and dyke injection that occur along ocean ridge axes.

Project FAMOUS (*F*rench-*A*merican *M*id-*O*cean *U*ndersea *S*tudy) made use of several of the navigation and survey techniques described in Chapter 1, first to select the area of study and then to examine it in great detail.

The FAMOUS area

The project started in 1971, when the theory of sea-floor spreading and plate tectonics had become accepted by most Earth scientists and the time was right for a more detailed survey of the areas where plates are generated—the oceanic ridges. The area chosen for project FAMOUS was a small segment of the Mid-Atlantic Ridge some 640 km south-west of the Azores (Figure 4.5 *inset*) at the boundary between the African and American plates. There were two main reasons for this choice: an aeromagnetic survey had revealed a clear central magnetic anomaly, and its position only a short distance from the port of Ponta Delgada in the Azores facilitated logistics, especially for the use of manned submersibles to explore the ocean floor.

Once the area had been chosen, a series of bathymetric surveys was carried out at progressively larger scales to produce maps such as those in Figures 4.5 and 4.6. On the small-scale map (Figure 4.5) you can see the

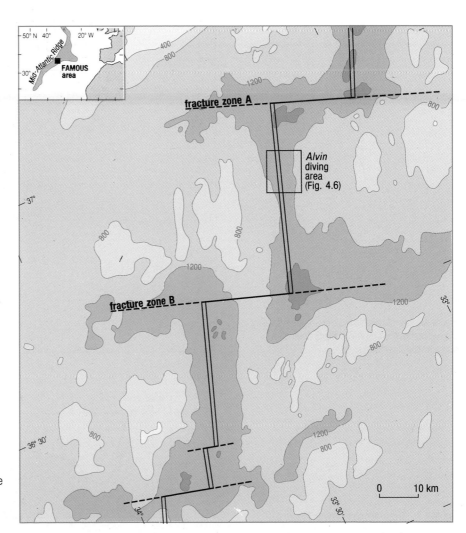

Figure 4.5 Map of the median valley on the Mid-Atlantic Ridge in the region of the FAMOUS area. The area of the submersible dives (Figure 4.6) is shown by the 'box' between fracture zones A and B. The contours are marked in fathoms.

Mid-Atlantic Ridge offset by a number of transform faults. Fracture zones A and B are about 50km apart and are the limits of the area covered by project FAMOUS. The area of the submersible dives is shown in more detail in Figure 4.6. The inner rift-valley floor is at a depth of about 2500m, and contains a number of small hills, of which the most prominent have been named Mount Venus and Mount Pluto.

Figure 4.7 is a cross-section through Mount Venus (Figure 4.6), crossing the median rift valley.

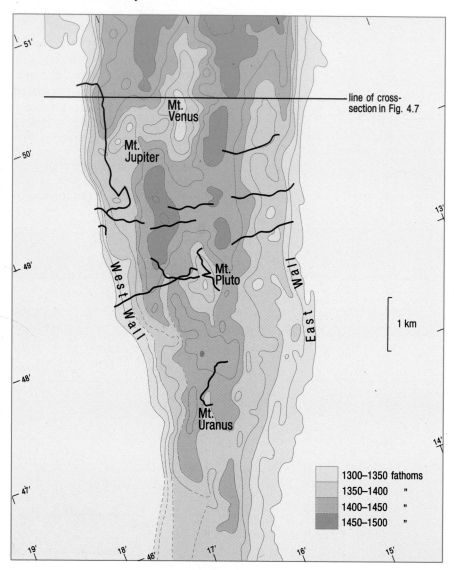

Figure 4.6   A more-detailed map of the rift valley in the region covered by *Alvin*. Black lines show submersible dive tracks.

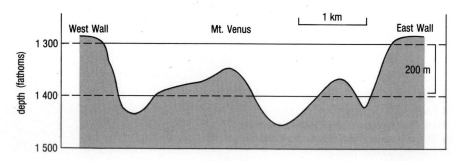

Figure 4.7   Cross-section through Figure 4.6.

QUESTION 4.3   (a)   Where would you place the largest faults on the rift-valley floor on Figure 4.7?

(b)   Where would you expect volcanoes on the rift-valley floor to occur on Figure 4.7?

Techniques

Almost every marine geological and geophysical technique which was feasible at the time was used during project FAMOUS. Many were techniques not normally used in marine studies at that time, and had to be devised or modified for the project.

1   Navigation system   As you know from Chapter 1, marine studies are of little use without some means of pinpointing the position at which measurements are taken. In the FAMOUS area, navigational problems were especially acute because of the extremely high precision required (to within about 10m). These problems were solved by developing an acoustic navigation system that made use of acoustic transponders placed on the sea-floor and 'linked' to the surface support ship and the submersible. This equipment proved so adaptable that it could be used with other instrument packages that were towed near the sea-floor.

2   Side-scan sonar   The British contribution to the FAMOUS project included a side-scan echo-sounding system known as *GLORIA* (Chapter 1). This gave images of the topography of the sea-floor. By making a mosaic of these records, topographic pictures of the sea-floor could be built up (Figure 4.8). These proved particularly useful in identifying major fault scarps and in delimiting the exact extent of the fracture zones.

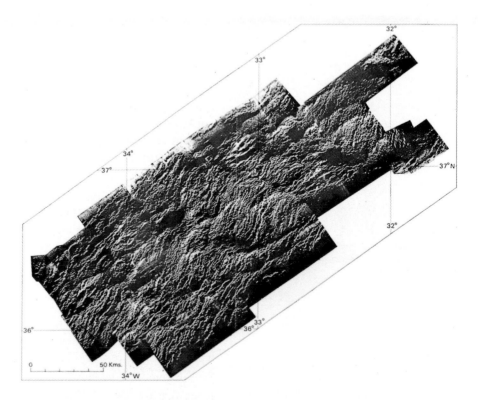

Figure 4.8   A *GLORIA* side-scan sonar mosaic of the FAMOUS area, obtained by combining overlapping sonographs generated by *GLORIA* scanning to the south-east. Areas which reflect the sonar beam strongly appear as white.

3 LIBEC The field-of-view and quality of conventional underwater photographs had always been limited by light scattering back towards the camera. This was overcome by using a high intensity *L*ight *Be*hind *C*amera system, towed above the sea-floor and designed to give large-area pictures. By taking a large number of photographs in rapid succession it proved possible to build up a photomosaic of the sea-floor. Figure 4.9 shows a typical mosaic, revealing some of the fissures characteristic of rift-valley floors.

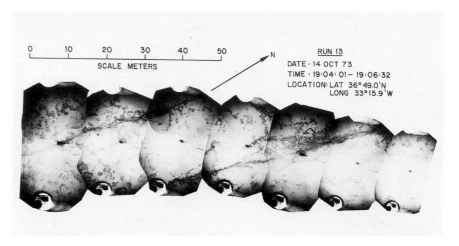

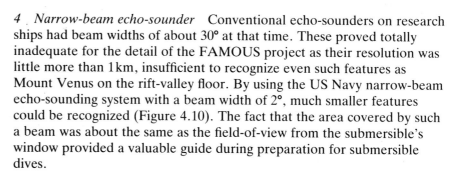

Figure 4.9 A *LIBEC* mosaic of the rift-valley floor, showing two of the numerous fissures that run parallel to the spreading axis. Note the much larger scale of this, compared with the *GLORIA* mosaic in Figure 4.8.

4 Narrow-beam echo-sounder Conventional echo-sounders on research ships had beam widths of about 30° at that time. These proved totally inadequate for the detail of the FAMOUS project as their resolution was little more than 1 km, insufficient to recognize even such features as Mount Venus on the rift-valley floor. By using the US Navy narrow-beam echo-sounding system with a beam width of 2°, much smaller features could be recognized (Figure 4.10). The fact that the area covered by such a beam was about the same as the field-of-view from the submersible's window provided a valuable guide during preparation for submersible dives.

These techniques paved the way for the manned submersible surveys. The submersible (Figure 1.21) had to be modified for work under high water pressures, and for this a high-pressure titanium hull was designed. Various sampling devices were also fitted, including a one-arm manipulator which could pick up specimens from the sea-floor, bags to collect loose sediment, plastic tubes to core sediment, and water-sampling bottles.

QUESTION 4.4 Figure 4.11 shows some of the photographs taken from *Alvin* on the rift-valley floor. For each photograph, which of the following features is/are present? (There may be more than one in each case.)

Well-formed pillow lavas.

Rock debris.

Sedimentary cover.

Fissures.

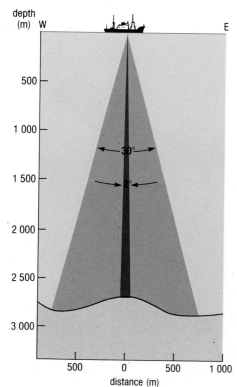

Figure 4.10 The value of using a narrow beam (dark red) rather than a conventional echo-sounder (light red) in making a topographic map of the rift-valley floor lies in its ability to resolve small topographic features such as the hill shown, which is about the size of Mount Venus (Figure 4.6).

(a)

(b)

(c)

(d)

Figure 4.11(a)–(d)   Some photographs of the rift-valley floor taken from the window of *Alvin*. The field-of-view of each photograph is about 1 m across for (a), about 3 m across for (b) and (d), and about 6 m across for (c). For details of (a)–(d) see Question 4.4 and its answer.

Figure 4.12   A section through a 50 cm diameter pillow lava recovered from the FAMOUS area.

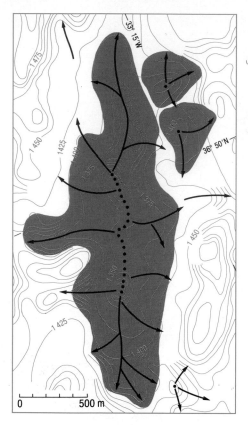

Figure 4.13 Topographic analysis of the bathymetric chart of the Mount Venus area. The red area is the high ground corresponding to Mount Venus. Dots are volcanic vents and arrows are flow directions deduced from submersible dives. Depths are given in fathoms.

Figure 4.12 shows part of an elongate pillow, which probably took only a few minutes to form from molten basaltic lava. The outer portion is made of glass formed when the surface of the lava flow was quenched very rapidly. Such glass quickly becomes hydrated, devitrified and altered by reaction with seawater. The interior cooled much more slowly (over a period of several days), which caused it to develop radial jointing. At the centre is a small tube or conduit that passes right through the pillow. The last portion of the lava to remain molten drained out along this conduit, probably feeding a new pillow bud.

Interpretation of the data
Analysis of the FAMOUS data began with the assumption that most features on the topographic map (Figure 4.6) were the result of volcanism. If this were the case, then each hill-like feature should have volcanic vents at or near the top and flow directions radiating from it. This was vindicated by detailed submersible observations on Mount Pluto (Figure 4.6), on which several vents were identified and the down-slope flow directions were confirmed by the flow morphology. Eighteen similar hill-like features occur on the rift valley floor within the FAMOUS area, all believed to be comparable in origin with Mount Pluto. The ones near

Figure 4.14 Sketch map of fault patterns and volcanic bodies in the inner rift valley of the FAMOUS area. Red shading indicates the youngest volcanism on the present rift axis. Heavy fault lines indicate wall faults; light fault lines, inner-floor faults. Black dots indicate vents and crest lines of volcanoes, and arrows show flow directions. Data are from bathymetric mapping, dive observations, bottom photography, and deep-tow geophysical studies.

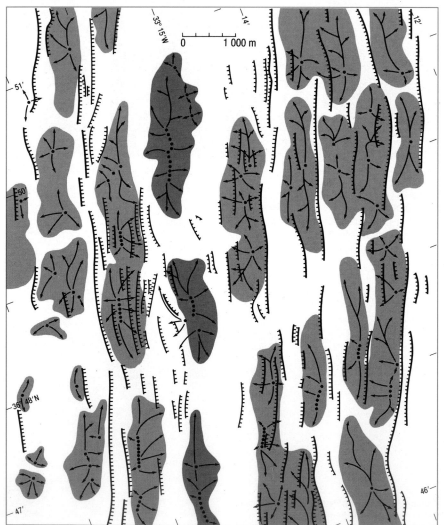

the centre of the rift, such as Mount Venus (Figure 4.13), seem to be the most recently active. Those further from the axis are older and are separated and occasionally cut by faults parallel to the axis (Figure 4.14).

Evolution of the rift-valley floor

Figure 4.14 gives a good idea of how complex the topography along the Mid-Atlantic Ridge is at the present time. Careful examination of the topography and identification of successive volcanic episodes enabled the FAMOUS team to build up the sequential diagrams that are displayed in Figures 4.15 and 4.16. These diagrams portray events that have taken place during the last 200 000 years or so. They represent the pattern of volcanism and rifting that has characterized the Mid-Atlantic Ridge for many tens of millions of years, since the Atlantic first opened. You can see successive stages in the formation of 2–3-km-wide strips of new oceanic crust either side of the Mid-Atlantic Ridge during the last 200 000 years, which is consistent with a (half-) spreading rate of 1–2cm per year.

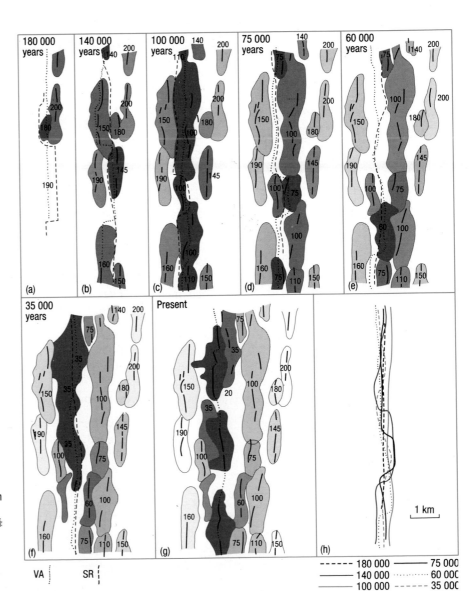

Figure 4.15   Evolution of the inner rift valley during the past 200 000 years, based on age estimates of the lavas. Dates of volcanic bodies are given as 10^3 years. For each stage (a)–(g), the active volcanic axis (VA) and the subsequent rift (SR) are shown. (h) All rifts superimposed. Note that transient offsets must have occurred at certain times; note also the coincidence between the SRs and VAs. See text for discussion.

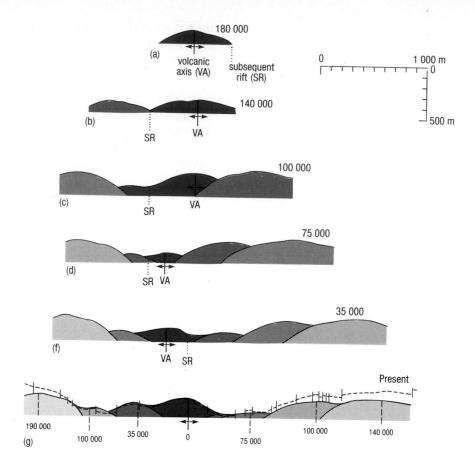

Figure 4.16 Cross-sectional diagram of the evolution of the inner valley, based on profiles across the valley floor after the elimination of faulting. The present faulted profile is shown in (g). Numbers indicate ages in years. (a)–(g) correspond to the stages in Figure 4.15. (*Note*: Stage (e) is omitted.)

You do not need to study all the details of these diagrams, for we summarize below the main points in the history of this particular piece of oceanic crust. They are a fair representation of what must be going on at other spreading ridges, albeit at different rates.

1   The frequency of volcanic events varies appreciably, the intervals ranging from about 10000 years to about 50000 years. Each volcano was formed by eruptions taking place over a period of several months to a few years. By the time it was built up, each volcano would be underlain by a swarm of dykes emanating from the magma chamber beneath.

2   After the volcanoes formed, they evidently subsided and rifting occurred. Rifts mostly form beside the volcanoes rather than splitting them, so the volcanoes are in general isolated as intact units. The rift may develop either to the east or the west of each volcano. At 180000 years ago, it lay to the west of the most northerly volcano, for example, but to the east of the more southerly ones. Each new phase of volcano-building develops about the latest rift. The new volcanoes, and especially their underlying dyke swarms, take up space, so the older volcanoes are moved aside to accommodate them.

3   By the time they are about 100000 years old, the volcanic hills are sufficiently far from the axis to begin to be uplifted by faults which mostly

dip inwards from the walls. After about 200000 years, the volcanoes have been upfaulted completely out of the valley floor on to the tops of the walls.

Clearly, the oceanic crust exposed outside the rift valley should equally consist of discrete volcanic units, also affected by faulting, which helps to account for the rugged and varied topography to be found along spreading ridges (Figure 2.12).

Surveys in other areas have shown that the processes revealed by the FAMOUS study and summarized above occur throughout the ocean ridge system, with differences of scale and timing as a consequence of contrasted spreading rates. Perhaps the most striking revelation of the FAMOUS study was that active axial volcanism, and presumably the dykes which feed the lavas, occurs in localized zones, and that there are intervals of local extension without volcanism between the extinction of one volcano and the growth of its neighbour. The frequency of dyke injection which you calculated in Question 4.2 is clearly an average valid only on time-scales of the order of 100000 years and longer. Even in two dimensions, then, the processes at ocean-ridge crests are more complicated than suggested in the cross-sections of Figure 4.3. You will see in Section 4.2 that further complexities become apparent when we consider the third dimension and examine changes in the pattern of activity *along* ridge axes.

4.1.3 MAGMA CHAMBERS AT SPREADING AXES

Volcanic processes are only the surface manifestation of the more deep-seated processes of oceanic lithosphere generation. If we want to understand how the plates are formed, we must look deeper. You should recall that seismic layer 3 is believed to be gabbro. This rock type is simply basalt that has been able to cool more slowly and form bigger crystals than in the pillow lavas and dykes. One reason for this is that it is deeper down, where the temperature is higher. Also, most geologists argue that, in order to cool slowly enough to produce crystals of the size observed in samples retrieved by dredging and drilling, this gabbro must have crystallized within a large chamber running along the axis (Figure 4.3(b)). Such a chamber would be kept hot by the frequent injection of new magma at the axis, and magma would be expected slowly to cool and crystallize at the edges of the chamber. If the crystallization at the chamber walls keeps pace with the injection of new magma from below, then the magma chamber must maintain a constant size.

A debate is continuing about how wide axial magma chambers are. They are likely to be wider at fast-spreading axes than at slow-spreading axes, and the width of 10km implied in Figure 4.3(b) is probably about the maximum that occurs. No active magma chamber has been positively identified on the axis of the Mid-Atlantic Ridge, but seismic studies on the axis of the East Pacific Rise have detected reflections believed to have come from the top of a magma chamber about 4–6km wide.

We will consider later how magma chambers may be supplied with magma from below, but first we should look at how processes along spreading axes vary, and in particular at the likely length of magma chambers.

4.2 SEGMENTATION OF OCEANIC SPREADING AXES

Studies in the FAMOUS area made it clear that the volcanic processes in that region are localized at discrete lengths of the axis. Each volcano is no more than about 2–3 km long and is separated from its neighbour along the axis by an inactive gap of almost 1 km (Figure 4.14). Does this mean that magma chambers are similarly spaced along the axis, with each volcano being fed from a different chamber? The answer is almost certainly no, because the geochemistry of the lavas from different volcanoes is too similar and the structure of seismic layer 3 argues against it. It is more likely that neighbouring volcanoes along the axis are likely to be fed from a single, much more elongated, magma chamber.

However, it is unlikely that a single magma chamber can run along the whole length of an ocean ridge axis.

Can you think why?

The obvious reason is because of transform faults. You have seen how transform faults divide ridges into segments that are between several tens and hundreds of kilometres in length. The offset on transform faults can be less than 20 km (e.g. the transform faults associated with fracture zones A and B in Figure 4.4) or as much as 1000 km (e.g. the Eltanin transform fault in Figure 2.23(d). It is difficult to see how a magma chamber less than 10 km wide could be continuous across such offsets.

Further evidence for the axial discontinuity of magma chambers comes from seismic reflection studies performed off-axis in the Atlantic. These suggest that the gabbro layer is thickest between fracture zones and may thin down to nothing at fracture zones. The only crust at many fracture zones appears to be dykes that propagated beyond the end of the magma chamber, and lavas fed by these dykes. This situation is illustrated in Figure 4.17.

Thus, axial magma chambers commonly do not extend right up to transform faults, and where they do the scale of the offset means that they are unlikely to be in contact with a similar chamber beyond the fault. Fundamental axial processes are thus segmented at least on a scale equivalent to that of transform fault spacing. During the mid-1980s, improvements in navigational accuracy and the deployment of narrow-

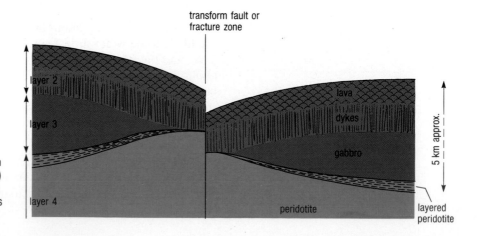

Figure 4.17 Schematic cross-section parallel to a spreading axis based on off-axis seismic reflection studies, showing how the rocks which crystallized in the magma chamber (gabbro and layered peridotite) get much thinner, and may disappear entirely near a transform fault. This indicates that magma chambers do not usually extend right up to transform faults.

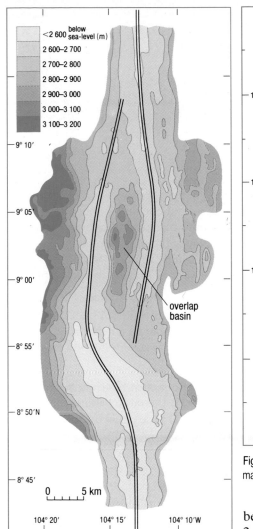

(a)

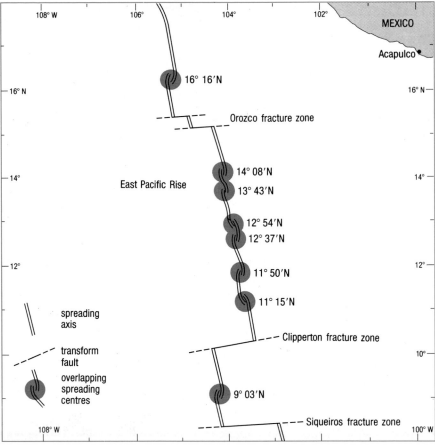

Figure 4.18   Overlapping spreading centres recognized along a 1000-km length of the East Pacific Rise mapped by narrow-beam echo-sounding and other observations.

beam echo-sounders and side-scan sonars revealed details of offsets of 2–15 km where axes have a 'kink', but there is no transform fault. Such non-transform offsets are becoming increasingly well known on the East Pacific Rise, where the straight segments between the kinks are several tens of kilometres long. You should recall from Figure 2.11(b) that the East Pacific Rise has no large median valley. Detailed bathymetric surveying has shown that the offsets on the axis occur where the axial crest begins to die away and a new axial crest begins to build up beside it. The places where axial crests overlap have therefore been named **overlapping spreading centres**. Figure 4.18 (above) shows eight overlapping spreading centres that have been mapped along a 1000 km length of the East Pacific Rise, which includes only four transform faults. The results of a narrow-beam echo-sounder survey of an overlapping spreading centre are reproduced in Figure 4.19(a) and (b).

Figure 4.19(a)   Bathymetric contours of an overlapping spreading centre on the East Pacific Rise. These were derived from narrow-beam echo-sounding on tracks separated by about 3 km. The axial crests are marked by double lines. Note that the area surveyed is considerably greater than the FAMOUS area shown in Figure 4.6.

(b)   Topographic cross-sections through the same overlapping spreading centre. The shallowest areas (less than 2700 m depth) are shown in black. These tend to consist of the youngest lava flows. Vertical exaggeration ×4.

(b)

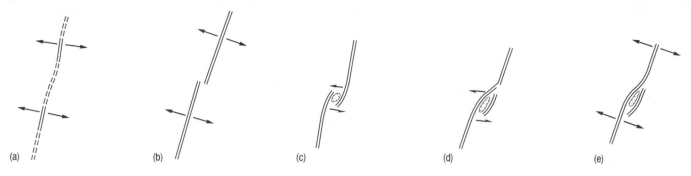

Figure 4.20   Possible model for the formation and evolution of an overlapping spreading centre. The double lines show axial crests above zones of active dyke injection.

(a)   Start of the cycle: ridge activity has been subdued for some time, due to lack of supply of magma into the chambers.

(b)   One or both magma chambers grow along the axis, due to the arrival of a fresh supply of magma, which causes the zones of dyke injection to propagate.

(c)   Axial crests (and zones of dyke injection) overlap, and begin to curve towards one another as a result of the stress field.

(d)   The more-vigorous zone propagates and cuts off the overlapping portion of the other one. At this stage, the two magma chambers may have merged.

(e)   The old overlap tip is abandoned, but remains as a topographic high. This is the end of a cycle. If magma supply wanes temporarily, the situation may return to (a). A complete cycle takes around 100 000 years.

Submersible observations show that the youngest pillow lavas occur on or very close to the axial crests, and it is clear that each crest must lie directly over a zone of dyke injection. Both volcanic and hydrothermal activity (see Chapter 5) are more intense near the middle of axial segments and die out towards the overlap zones, whereas fracturing and small-scale faulting increase towards the overlap, presumably in response to stresses induced by the offset.

It is not yet clear whether the axial segments either side of overlapping spreading centres are being fed from separate magma chambers. However, many of them occur at places where seismic reflection from the supposed magma chamber roof dies out.

Thus, it appears that the axial magma chambers must have extra breaks apart from those at transform faults. In fact, if the model shown in Figure 4.20 is correct, the breaks in the magma chamber are responsible for the very existence of the overlapping spreading centres.

QUESTION 4.5   (a)   What are the approximate average lengths of axis segments on the Mid-Atlantic Ridge in the FAMOUS area (Figure 4.5) and on the East Pacific Rise between 8°N and 18°N (Figure 4.18)?

(b)   Does this suggest any correlation between spreading rate and the scale of segmentation?

4.2.1   A PLAUSIBLE MODEL FOR LITHOSPHERIC GROWTH

It was suggested in Section 4.1 (Figure 4.3) that the oceanic crust is produced by partial melting of asthenospheric mantle which rises beneath spreading axes.

But how can the rising asthenosphere produce regularly spaced magma chambers of the kind required to cause segmentation?

The lithosphere is denser than the partially molten asthenosphere; over geological time-scales this situation is analogous to that where a layer of dense fluid overlies a less dense one. It has been shown experimentally that if the density and viscosity contrasts across such a boundary are right, then the less dense layer will rise into the upper layer in the form of regularly spaced protrusions (Figure 4.21(a)).

At a spreading axis, the lower layer is a linear zone of lower density asthenosphere beneath the ridge (cf. Figure 4.3), but the effect should be the same. Figure 4.21(b) is an idealized diagram illustrating how magma concentrations within the asthenosphere should naturally rise to form

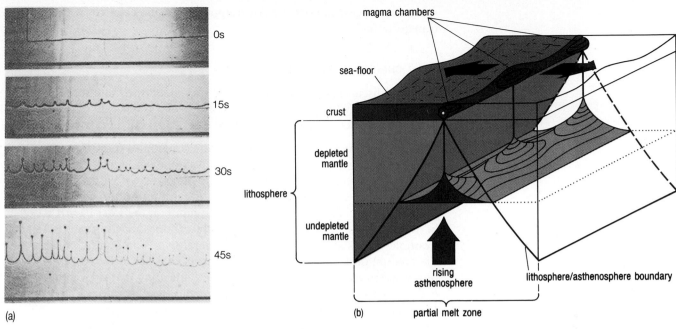

Figure 4.21(a) Gravitational instability of a water–glycerine mixture (less dense) injected rapidly along a horizontal line into glycerine (more dense). A fairly regular spacing of rising 'blobs' of the less dense mixture is apparent after 45 seconds. The water–glycerine mixture represents partial mantle melts, and the individual blobs represent feeder zones for individual magma chambers in (b).

(b) Viscous asthenosphere rises two-dimensionally beneath the boundary between two spreading lithospheric plates. Above a certain level (dotted lines), the rising asthenosphere passes into a zone in which partial melt can form, and collect at some level below the base of the lithosphere. Due to its lower viscosity and density, the partial-melt zone develops a gravitational instability leading to regularly spaced concentrations of melt which percolate towards the surface to form discrete crustal magma chambers giving rise to segmentation along spreading axes. The asthenosphere continues to rise viscously and on cooling becomes rigid and joins the lithosphere which thickens away from the spreading axis. The bold lines mark the boundary between lithosphere and asthenosphere.

magma chambers beneath individual spreading segments. The spaces between the segments will be starved of magma, topographically lower, and have thinner crust. If the axis is offset sufficiently, a transform fault will form.

Once a magma chamber has been formed in this way, it is likely to continue to be fed from the same source, because the gravitational instability pattern in the zone of asthenospheric partial melting will probably persist. However, you will see in Section 5.3.1 that there is strong evidence that axial magma chambers may often be sufficiently cooled by hydrothermal circulation to crystallize completely, so that they are episodic in time as well as discontinuous in length.

In summary then, a spreading axis is fed by magma produced by partial melting of asthenosphere in an upwelling zone along the axis. Magma chambers are discontinuous along the axis, on a scale of several tens of kilometres, and each magma chamber may supply lava to several volcanic vents situated every few kilometres along the axis. This model is supported primarily by detailed morphological surveys and seismological studies. Geochemical work reveals subtle but distinct contrasts in the composition of basalts from different ridge segments.

Where an axis is offset by more than about 20km there is almost always a transform fault, but smaller amounts of offset, particularly on fast-spreading ridges, can occur as overlapping spreading centres without transform faults.

4.2.2 CHANGES IN SPREADING PATTERN

We have discussed how spreading axes vary along their length, and how overlapping spreading centres may evolve over a time-scale of 100000 years. What about longer-term changes?

Magnetic anomaly patterns and other data sometimes show that a segment of ridge axis has jumped sideways by several tens or even a few hundreds of kilometres. Sometimes ridge jumps occur by the propagation of the end of a segment across a transform fault, but often a ridge jump has no obvious cause. The possible ridge jump on the East Pacific Rise between the Menard and Udintsev Fracture Zones (Section 2.6) is a good example. You should be able to find several places on Figure 3.6 where the amount of displacement between ocean floor age strips on either side of a fracture zone has changed over time, indicating that ridge jumps have occurred. The anomalously wide age strips to the east of the East Pacific Rise between 10°S and 35°S are probably due to a large ridge jump. In this case, the old ridge is still visible as a bathymetric feature between the active East Pacific Rise and the coast of northern South America (Figure 1.11).

Spreading axes must also adjust whenever the direction of plate motion changes. Sharp changes in spreading direction of up to about 10° at a time occur every few million years and are an inevitable geometric consequence of having several plates jostling against one another on the surface of a sphere (the Earth). This can be seen from the changes in direction of fracture zones and sea-floor age strips. There are several examples of this in the Pacific, in particular.

The most major change in ridge axis configuration occurs when a totally new ridge forms. The development of the South-west Indian Ridge is an example we have already come across (Section 3.3).

4.2.3 CRUSTAL ABNORMALITIES

Fully developed normal oceanic crust (seismic layers 2 and 3) which formed at a major ridge axis is 7km thick on average, and the relative proportions of different crustal layers approximate closely to what you saw in Figure 4.1. However, oceanic crust that formed at spreading axes in back-arc basins (*cf*. Figure 3.8), while having the same structure as normal oceanic crust, is generally 2–3km thinner.

In some areas, by contrast, abnormal thickening of oceanic crust has occurred. The best-known example is the island of Iceland which lies on the Mid-Atlantic Ridge. Here the processes of crustal generation can be seen on land—indeed, the study of volcanic and structural processes in Iceland has done much to confirm and refine the concepts of sea-floor spreading and plate tectonics. A regular spacing of volcanic centres has also been observed here, consistent with the segmentation model outlined in Section 4.2, and lateral shifts of the main spreading axis are well-documented (Figure 4.22). Crust which is oceanic in structure but approaches the thickness of continental crust has been found in regions as far apart as the Arctic and the western Pacific, but these occurrences are less well-documented than Iceland because they lie below sea-level.

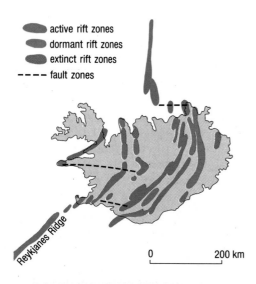

active rift zones
dormant rift zones
extinct rift zones
- - - - fault zones

Figure 4.22 Map of Iceland showing its position on the Mid-Atlantic Ridge and the distribution of active rift zones. Note how the spreading axis (defined by rift zones) has changed position with time. The Reykjanes Ridge is part of the Mid-Atlantic Ridge system.

0   200 km

There is also considerable variation in seismic layer 1, which does not in fact consist exclusively of sediments. Drilling projects such as ODP and DSDP (Section 1.3) have revealed that igneous rocks occur quite commonly within the sedimentary sequence. We know that volcanism forming seamounts and guyots can be much younger than the underlying igneous rocks upon which they are built (Section 2.5.2). It follows that layers of lava may be interlayered with the sediments. For example, the Nauru Basin, east of the Marianas Trench in the western Pacific, is floored by crust of Jurassic age (*c*. 150 Ma old), but there was a major episode of volcanism unrelated to plate boundaries in this region during the Cretaceous, 80–90 Ma ago. This did not form any seamounts or volcanic islands in the area, but instead produced lava flows among the sediments, and some sheets of basalt were emplaced along bedding planes as sheet-like igneous intrusions (sills).

In the early phases of ocean basin formation (Stage 2, Table 3.1), sills are commonly emplaced within the rapidly accumulating sedimentary pile of the subsiding rift (*cf*. Figure 3.2). The Gulf of California, where a substantial sill–sediment complex has been found, is a good example.

Where there is much faulting and fracturing, as in axial rifts and along transform faults and fracture zones, deeper layers of normal oceanic crust can be exposed at the sea-bed. The greatest vertical displacements are typically at ridge–transform intersections.

Indeed, many samples of the full range of rock types representing the four layers shown in Figure 4.1 had been dredged from the ocean floors long before the realities of sea-floor spreading were recognized, and provided key evidence for our understanding of oceanic crustal structure. Nowadays, submersible and deep-sea camera surveys, as well as drilling operations, frequently reveal rocks typical of layer 3 (gabbro) and layer 4 (peridotite) exposed in the walls of median valleys or the scarps and cliffs of transform faults.

The interaction of seawater with peridotite of layer 4 alters it into a rock known as **serpentinite**, which is much less dense than peridotite and more plastic, so it can be squeezed up towards the surface along fractures in the crust. This occurs especially at 'leaky' transforms (Figure 2.17) where it is common to find extrusions and intrusions of serpentinite, and indeed other igneous rock types.

In short, although the structure of oceanic crust is very simple and straightforward in general, it can be locally complicated in detail. It is now much easier to detect and investigate these local complications than it used to be. Sophisticated geophysical instruments of the kind widely used in the oil industry are lowered down boreholes to measure directly the properties of the rocks. Parameters such as fracture spacing, clay mineral content, seismic velocity and degree of metamorphic alteration can all be measured or deduced, and direct observation is possible with down-hole video cameras. Careful calibration of the instruments, using cores obtained from selected reference boreholes, has made it possible to dispense with the very expensive business of obtaining rock cores from every borehole drilled into igneous oceanic crust.

So far, we have concentrated mainly on processes that occur at ocean ridges, but while these may be the most active volcanic regions on Earth they are not the only regions of volcanism in the ocean basins.

4.3 SEAMOUNTS AND VOLCANIC ISLANDS

A glance at Figure 1.11 shows that seamounts and volcanic islands are a ubiquitous feature of all the oceans, though they are more abundant in some parts (e.g. the Pacific) than others (e.g. the Atlantic). Submarine volcanoes have shapes that are generally similar to those of some land-based volcanoes (Figure 2.20), and they grow in much the same way—by successive additions of lavas (in this case mainly pillow lavas) and ashes.

Volcanic ashes are very rare in the oceanic crust proper, but merit some attention at this point. Water depths of a few kilometres exert hydrostatic pressures in the order of a few hundred atmospheres, enough to prevent dissolved gases escaping from the lavas. However, at depths of less than about 500 m, gas pressure may well overcome the hydrostatic pressure, and the rapid escape of dissolved gases can completely fragment the lava to form a volcanic ash consisting of glassy lava fragments. The lower parts of oceanic volcanoes are therefore largely made up of pillow lavas, but the upper parts consist of pillow lavas with ashes if the volcano built up sufficiently close to sea-level to allow gaseous fragmentation of the lava.

At shallow depths, the chances of explosive eruption and fragmentation of lava are also enhanced when seawater comes into contact with enough hot magma to raise its temperature rapidly to boiling point—the resulting sudden expansion to steam may produce spectacular explosions (Figure 4.23).

Most seamounts are volcanoes that never grew to sea-level; others stopped erupting above sea-level and were subsequently eroded.

What explanation have we offered for the flat tops of guyots that may now be hundreds of metres below sea-level?

They were planed off near sea-level by the action of waves (Section 2.5.2) and have since subsided along with the cooling lithospheric plate beneath them (*cf*. Figure 2.13).

Figure 4.23 The explosive eruption of Surtsey (south-west of Iceland), November, 1963. When submarine volcanoes build up to near sea-level, and seawater comes into contact with hot magma, the confining pressures are no longer sufficient to prevent explosive conversion to steam. After the volcano has built up far enough above sea-level for the magma column to be isolated from seawater, this type of eruption ceases. Volcanoes of course commonly explode for other reasons, such as the build up of gas pressure in the magma.

However, the flat tops of some guyots could be congenital rather than developed after formation of the volcano. Figure 4.24(a) shows a flat-topped volcano thought to have been formed below sea-level and subsequently uplifted. This occurred in the Afar region of Ethiopia, at the southern end of the Red Sea, a region well-known for tectonic instability. Figure 4.24(b) shows a steep-sided and relatively flat-topped seamount in the eastern Pacific. Narrow-beam echo-sounding and side-scan sonar techniques have revealed that the 'flat' top is due to the presence of a large, shallow crater.

Figure 4.24(a)   Asmara volcano, west of Lake Abbe (central Afar). This truncated cone does not result from the levelling of some supposed earlier and more pointed cone but is actually congenital.

(b)   Side-scan sonar (*Sea MARC*) image of a seamount on the west side of the East Pacific Rise at 9°55′N. The summit crater is about 2 km across, and its floor is about 100 m below the rim. This crater is presumed to have been formed by collapse as magma was withdrawn from below. There are smaller craters and other volcanic features within it. The highest point on the seamount is about 1640 m below sea-level, and it stands about 1 km above the sea-floor. As it is only about 330 000 years old, the general flat-topped nature is unlikely to be due to processes that occurred near sea-level. Colours are superimposed to show depths.

(a)

(b)

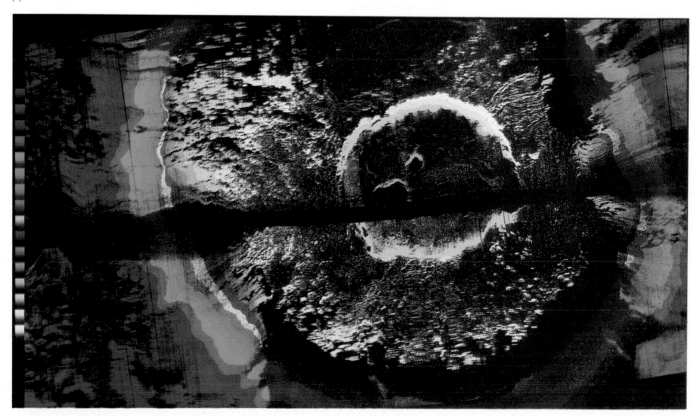

This Chapter had the twin objectives of (i) giving you an insight into the kinds of detailed information that can be obtained by the application of sophisticated techniques in sea-bed exploration, and (ii) of providing you with a background for studying the processes of hydrothermal circulation which occur mainly at ocean-ridge crests—the subject of the next Chapter.

4.4 SUMMARY OF CHAPTER 4

1   Oceanic crust has a well-defined layered structure. Layer 1 consists mostly of sediments overlying the igneous crust of layers 2 and 3; layer 4 being the uppermost (lithospheric) mantle. Layer 2 is volcanic, dominated by pillow lavas (above) and dykes (below). Layer 3 is gabbro and represents the solidified contents of sub-ridge magma chambers. The total thickness of normal igneous oceanic crust is about 7km. In both chemical and mineralogical composition, virtually all the rocks of layers 2 and 3 are basaltic. Layer 4 is of peridotite composition. The principal seismic discontinuities in oceanic crust are at the base of layer 1 and the top of layer 4 (the Moho). Seismic velocities generally increase with depth, and variations in the gradient of this velocity change enable subdivisions of the crust to be recognized.

2   Several sophisticated techniques—including narrow-beam echo-sounding, side-scan sonar, underwater photography, and submersible operations—have shown that volcanism in the median rift valley of ocean ridges is not continuous but episodic, at intervals of around 10^4–10^5 years. Lavas are erupted from elongate volcanoes rather than from a continuous fissure, and the line of separation of lithospheric plates is irregular on the kilometre scale.

3   Spreading axes are segmented at intervals of about 30–100km. The segments are separated either by transform faults or by overlapping spreading centres. Ridges are highest, and the crust thickest, at the midpoints of segments, where the volcanic activity is greatest. This falls off towards the end of segments, where the ridge is topographically lower, and the crust thinner. The gabbro of layer 3 sometimes thins out to nothing close to fracture zones.

4   As a model for axial processes, we suggested that an elongate zone of less dense asthenosphere rising beneath denser lithosphere will form regularly spaced protrusions into the overlying layer. Each of these protrusions feeds a magma chamber beneath a spreading ridge segment. Each magma chamber may feed several volcanic vents in an episodic fashion.

5   Oceanic crust is thinner than normal in back-arc basins and thicker than normal in some other parts of the oceans. Fractures and faults can bring rocks of deeper crustal layers to the surface.

6   Seamounts and volcanic islands are formed by isolated submarine volcanoes, building up from the sea-bed. Only a minority reach the surface to form islands. Flat-topped seamounts (guyots) are mostly islands that have been planed off by wave action, but some may have formed with flat tops due to volcanic processes. Dissolved gases cannot escape from lavas at depths of more than about 500m, so volcanic ashes are rare in the deep oceans.

Now try the following questions to consolidate your understanding of this Chapter.

QUESTION 4.6 (a) Overlapping spreading centres have been described in the literature as: 'complex and variable and inherently unstable on length- and time-scales of about 10 km and 10^5–10^6 years respectively, though on longer length- and time-scales they appear to be stable'. Explain what this means.

(b) Look at Figure 4.19(b), which shows both 'limbs' of the overlapping spreading centre at around 9°N. From which magma chamber(s) would you expect rock samples dredged from each limb on the profile at 9°00.5′N to have come, and would you expect the samples to be similar or different in terms of their chemical and mineralogical composition?

QUESTION 4.7 Although some dissolved gas generally escapes from any lava, often some gas remains trapped in the rock in small, bubble-like cavities called *vesicles*, which are generally less than 1 cm in diameter. Other things being equal, would you expect vesicles in submarine lavas to be larger or smaller than those in terrestrial lavas? Does your answer help to explain why volcanic ash is rare in the lower parts of oceanic volcanoes?

CHAPTER 5

HYDROTHERMAL CIRCULATION IN OCEANIC CRUST

Figure 5.1 A 'black smoker' in the axial zone of the East Pacific Rise. Heated and chemically changed seawater emerges from the sea-bed as a clear fluid at about 350°C, and immediately precipitates metal sulphide particles on contact with cold bottom water, building the vent chimney and forming the dense plume of black 'smoke'. The vent is about 20 cm across.

The discovery of hot springs on the ocean floor during the 1970s was one of the most exciting events in the history of oceanography. The most spectacular are the **'black smokers'** (Figure 5.1) where hot water gushes out of the vents in the sea-bed at temperatures of 350°C or more, forming a dense plume of black 'smoke' made up of minute particles of metal sulphides. At lower vent temperatures (30–330°C), the 'smoke' is dominated by white particles of barium sulphate, giving rise to the term **'white smokers'**. Less spectacular but just as significant are the lower temperature **warm-water vents** where water emerges about 10–20°C above surrounding bottom-water temperatures of 2–3°C. These sources of hot water support unusual ecosystems (Figure 5.2), in which the primary production that underpins the local food web depends not on photosynthesis but on chemosynthetic bacteria, which derive their energy by the oxidation of sulphide from the vents.

Once the oceanic crust has been formed by igneous activity, hydrothermal processes take over, as seawater (driven by convection) circulates within the newly formed hot igneous rocks. This is by no means a minor phenomenon, and it is likely that about one-third of the entire sea-floor contains active water circulation systems. The rate of circulation is sufficient for every drop of the 1.4 billion (10^9) km³ of ocean water to pass through the oceanic crust in a few million years, leading to major exchanges of chemical elements between seawater and hot basalt. Consequently, the crust buffers the chemical composition of the oceans and for some elements in seawater it is a more important source than rivers. The formation of concentrations of metal sulphide which accompanies the discharge of hydrothermal fluids into the ocean is one of the Earth's principal mechanisms of ore generation. Sulphide ore deposits in ophiolite complexes such as those in Cyprus and Newfoundland are well-known examples.

It was predicted that hydrothermal circulation must occur in oceanic crust at ridge crests several years before the first vents were discovered. In the

Figure 5.2 Tube worms are a major part of the ecosystems round hydrothermal vents. The worms are red and can be seen protruding from their chitinous tubes, which may reach several metres in length. The worms have no stomachs, but absorb nourishment directly from the water. The small crabs in the picture are blind and have no eyes in their eye-sockets, which have become adapted into scrapers for scratching off micro-organisms coating the outside of the worm tubes, on which they feed.

Figure 5.3 Geyser activity in Iceland. Hot water and steam rise explosively and intermittently from small vents.

mid-1960s, the occurrence of hydrothermal systems in volcanic areas on land led to the proposition that similar systems should also be found along the ocean-ridge system, which had recently been recognized as a zone of active volcanism. The hot springs and geysers of Iceland (Figure 5.3), which straddles the Mid-Atlantic Ridge (Figure 4.22), provided obvious and highly visible evidence that hydrothermal activity could indeed occur at ridge crests. At about the same time, chemical analyses of samples of the most recently deposited sediments on the sea-floor revealed a systematic increase in the concentration of iron, manganese and some other metals (e.g. Ag, Cr, Pb, Zn) towards the ridge crests (Figure 5.4). It was clear that a local source must be responsible for this pattern, and hot spring activity at ridge crests provided the best explanation for the observed trends.

Further support came from samples of basaltic rocks dredged from ridge axes, many of which show clear evidence of alteration and metamorphism by reaction with hot seawater. Studies of ophiolite complexes and their associated ore deposits confirmed that large volumes of seawater can penetrate more than 5 km into oceanic crust and circulate within it at high temperatures.

Figure 5.4   A map (made in the 1960s) showing the proportions of Al, Fe and Mn in the uppermost sediments on the ocean floor. The ratio shown decreases with distance from ridge crests, because Al is a major constituent of the clay minerals that occur in all deep-sea sediments, and are supplied to the oceans mainly from continental weathering; whereas Fe and Mn are deposited primarily as a result of hydrothermal vent activity near ridges.

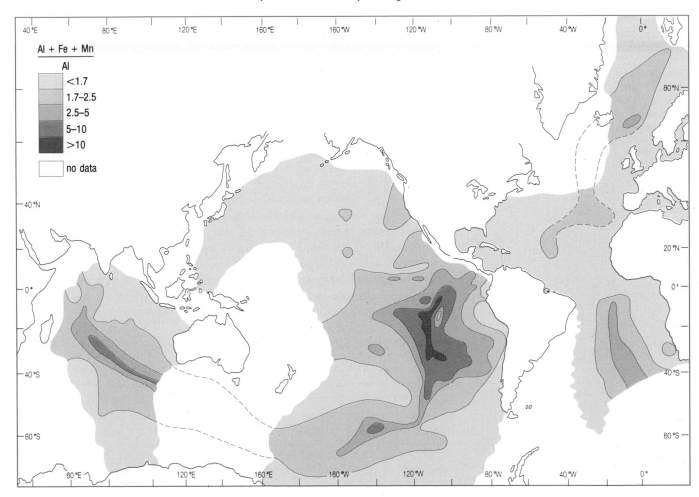

5.1 THE NATURE OF HYDROTHERMAL CIRCULATION

Hydrothermal systems have two basic characteristics: they occur in regions of high geothermal gradient where hot rocks lie near the surface, and they have a 'plumbing system' of fractures so that cold water can percolate downwards into the crust and then flow upwards to the surface as hot water. The movement of water through these fractures is such that downward percolation occurs over a wide area, through pores and cracks and crevices in the rocks, whereas the upward flow is concentrated through a limited number of channels, which is why it often escapes so vigorously (Figure 5.5).

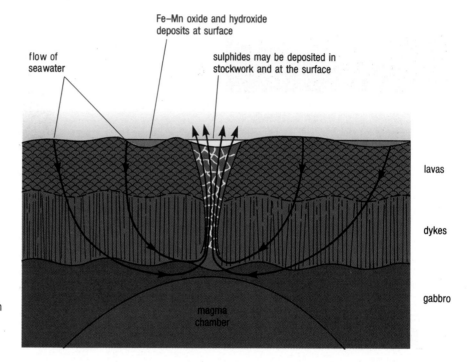

Figure 5.5 Schematic cross-section, normal to a spreading axis, to illustrate hydrothermal convection in oceanic crust. Flow lines indicate downward and lateral percolation of seawater and its expulsion upwards in a hot narrow plume. This can lead to the deposition of ore minerals on the surface and within the crust; hence the high concentrations of metal-rich sediments at ridge crests indicated by Figure 5.4. The term *stockwork* is explained later—Section 5.3.1.

The principal difference between oceanic and terrestrial hydrothermal systems is that the oceanic crust is overlain by thousands of metres of seawater and subject to high hydrostatic pressure, whereas on land there is only the atmospheric pressure. In global terms, oceanic hydrothermal activity is by far the more important of the two. It occurs along the whole length of the ocean-ridge system and for large distances on either side, and takes place continuously, because the generation of new oceanic crust is continuous on the geological time-scale. Temperatures are higher and flow rates greater than in any terrestrial hydrothermal system.

5.1.1 HEAT FLOW, CONVECTION AND PERMEABILITY

The oceanic crust is hot at its lower boundary whereas the upper boundary is in contact with seawater or wet sediments at a few degrees above 0°C. The temperature difference is particularly marked near active ridges, where hot magma chambers lie within the crust (at above 900°C, unless completely crystallized), and there is little or no sediment cover.

Heat must rise from the hot to the cold part of the crust. In the solid crust this can happen in two ways: by conduction, in which heat diffuses upwards; and by convection, in which the heat is transferred by the mass movement of water, i.e. by hydrothermal circulation. The question is, which of the two is more important?

The temperature difference between the top and bottom of the crust becomes progressively less as the lithosphere cools on moving away from the active ridge crest. You have already seen how thermal contraction of the lithosphere leads to an exponential relationship between age and depth to the sea-floor (Figure 2.13). Theoretically, a similar exponential relationship with age should hold for the rate of loss of heat from the top of the crust, if it is assumed that the heat is lost by conduction only, i.e. by upward diffusion through the solid crust.

Measurements of conductive heat flow (the rate of heat loss by conduction per unit area of surface) are made by embedding sensitive heat-measuring instruments into sediments on top of the igneous crust. Figure 5.6 encapsulates some of the results of both theoretical prediction and actual measurement throughout the ocean basins.

Figure 5.6   Theoretical and observed profiles for heat flow (rate of loss of heat per unit surface area) versus age of oceanic crust for different parts of the world's oceans. Dots at centres of crosses are regional averages of measured heat flow over the age intervals represented by horizontal bars. Vertical bars represent one standard deviation from the mean of the heat flow measurements. The dashed curves represent the theoretical exponential decay of heat flow away from mid-ocean ridges, assuming heat loss by conduction only. The shaded areas represent the 'missing' non-conductive heat loss, which is not measured by the instruments.

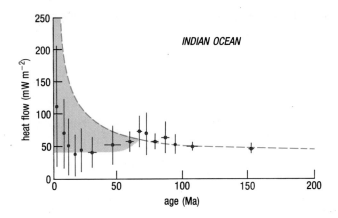

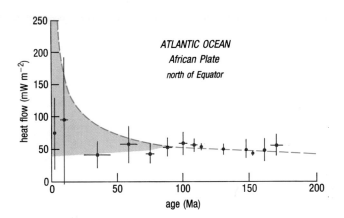

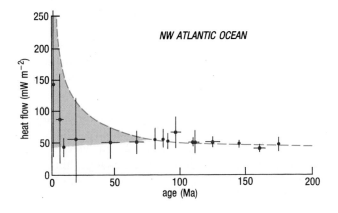

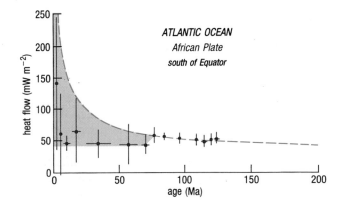

QUESTION 5.1 (a) Theoretical computations predict a large *conductive* thermal anomaly at ridge crests. Do the actual observations confirm this prediction?

(b) What is the significance of the shaded areas on Figure 5.6?

(c) How do the theoretical curves in Figure 5.6 resemble Figure 2.13?

(d) How does Figure 5.6 support the statement at the beginning of this Chapter that about one-third of the sea-floor has active hydrothermal systems operating in it?

Recognition of the 'heat loss deficit' you have identified in Question 5.1 provided marine scientists with the first unequivocal evidence that hydrothermal circulation through oceanic crust must occur on a very large scale; this was still in the early 1970s, some years before any vents were actually discovered. For convection to take place, the two most important conditions are that the thermal gradient must be high enough to overcome the various physical forces acting against fluid motion, and that there must be channels in the rocks for the water to move through—in other words, the rocks must be permeable.

What provides sufficient permeability in oceanic crust to enable seawater to circulate?

Major faults and fractures are obviously important, but there are also numerous smaller fractures in the rocks—especially the fractures in pillow lavas, the spaces between the pillows and among the 'rubble' of seismic layer 2A, and fractures within and between dykes.

Permeability is likely to be greatest near active ridges, where cooling of hot igneous rocks has caused new fracturing. As the crust moves away from the spreading axis, channels become clogged by minerals precipitated from the circulating fluids and the rocks are covered by progressively thicker, less permeable, sediments.

5.2 CHEMICAL CHANGES DURING HYDROTHERMAL CIRCULATION

In anticipation of direct observational proof that seawater convection occurs in oceanic crust, numerous laboratory experiments were made to simulate the likely conditions. Seawater was cycled through powdered basalt at elevated temperatures and appropriate pressures for varying lengths of time, using a range of **water:rock ratios**. Dramatic changes were observed, and the reactions were very rapid in terms of geological time. For example, over periods that ranged from weeks to months, virtually all the magnesium and sulphate in the seawater was transferred into the rock, while significant amounts of potassium, calcium and silicon were leached from the rock by the seawater.

It soon became clear that hydrothermal activity must have been a major and previously unconsidered contributor to the chemical mass balance of the oceans throughout the history of the Earth. Experiments were continued in order to try to quantify more precisely the changes that occur, as many components of the natural systems cannot be sampled easily.

5.2.1 CHANGES IN THE ROCKS

IMPORTANT: It is not essential to know or remember the identity of all the rocks and minerals named in this Section. If your experience of rocks and minerals is limited, all you need to understand is the following list (1–3) and then continue where indicated:

1   Rocks of basaltic composition have completely crystallized by the time they have cooled to 900°C or so—i.e. the rocks become solid at about that temperature—the exact temperature depending on pressure and water content. They consist of mixtures of mineral crystals (plus some volcanic glass in the case of pillow lavas) that are chemically unstable in the presence of seawater. Even cold seawater can cause chemical changes and the hotter the seawater, the more profound those changes become.

2   In cold seawater, basaltic rocks experience **sea-floor weathering**, which is similar to that which occurs on land, but they are transformed (metamorphosed) into different rock types by reaction with heated seawater during hydrothermal circulation.

3   Hydrothermal **metamorphism** most often occurs at temperatures and pressures such that the rocks are metamorphosed to **greenschists**. These then consist of mixtures of minerals that are different from those making up the original basalt. At higher temperatures and pressures, basalts in the deeper parts of seismic layer 2 and gabbros in the upper parts of layer 3 may be transformed to **amphibolites**. Sea-floor metamorphism changes the overall appearance of basaltic rocks very little. In fact, the specimens illustrated in Figure 4.2(b)–(e) are not fresh rocks but have been subject to sea-floor metamorphism and weathering.

To see the effects of these processes, we should consider the *mineralogy* of the rocks. The principal mineralogical changes that may be experienced by basaltic rocks on the sea-floor are summarized below.

(i)   The minerals that make up rocks of basaltic composition (i.e. the pillow lavas and dykes of seismic layer 2, and the gabbros of seismic layer 3 (Figure 4.2)), when they are fresh and unaltered, are principally:

calcium-rich plagioclase feldspar. $CaAl_2Si_2O_8$ (*c.* 50–70%):

pyroxene, $Ca(Mg,Fe)Si_2O_6$ (*c.* 30–40%);

olivine, $(Mg,Fe)_2SiO_4$ (*c.* 0–10%);

and glass of basaltic composition (up to 70% in some pillow lavas).

(ii)   At the sea-bed, the pillow lavas forming the top of seismic layer 2 are in contact with oceanic bottom water at temperatures of 2–3°C or less. Chemical reactions here come under the general heading of sea-floor weathering. These are similar to processes which occur when rocks are weathered on land, and involve the alteration of feldspars and glass to clay minerals (hydrated aluminosilicate compounds), as well as oxidation of iron and precipitation of manganese oxide as films and crusts. Sea-floor weathering occurs wherever basaltic rocks are exposed on the sea-bed.

(iii)   Minerals belonging to another group of hydrated aluminosilicates, called zeolites, may also be formed, both by sea-floor weathering, and where temperatures have been increased to 100–200°C during hydrothermal circulation.

(iv)   At temperatures in the range of around 200–400°C, which are typical of those in most oceanic hydrothermal systems, a wholly new assemblage of minerals is formed, comprising various combinations of the following:

>sodium-rich plagioclase feldspar (albite), $NaAlSi_3O_8$;
>
>chlorite (a dark green mica-like mineral), $(Mg,Fe,Al)_3(Si,Al)_2O_5(OH)_4$;
>
>quartz, SiO_2;
>
>epidote, $Ca_2(Al,Fe)_3Si_3O_{12}(OH)$;
>
>actinolite, $Ca_2(Mg,Fe)_5Si_8O_{22}(OH)_2$.

When this mineralogical assemblage has been reached, the rocks have become greenschists, which are the commonest metamorphic rocks of the oceanic crust. The proportion of minerals in the greenschists can vary greatly from place to place, depending on variables such as temperature gradient, duration of the activity, and relative volumes of water and rock; but they will all have formed within the temperature range of around 200–400°C and several hundred atmospheres pressure.

(v) Seawater sometimes penetrates to deeper levels in the oceanic crust where temperatures and pressures are still greater. Under these conditions, in rocks of basaltic composition (typically the lower parts of the dykes making up seismic layer 2C, upper parts of the gabbro making up layer 3), a new mineral—hornblende—appears:

>$(Na,Ca)_2(Mg,Fe,Al)_5(Si,Al)_8O_{22}(OH)_2$.

Hornblende-bearing rocks are known as amphibolites.

(vi) Finally, there are the serpentinite intrusions, referred to in Section 4.2.3. Seawater can penetrate to seismic layer 4 along transform faults and other major fractures. It hydrates the olivine which is the main constituent of peridotite, transforming it into serpentine, $Mg_3Si_2O_5(OH)_4$. The resulting rock is a serpentinite, which deforms plastically at moderate temperatures. Being less dense than rocks of the overlying seismic layers 2 and 3, serpentinite is often squeezed upwards along fractures in the crust.

Note: Readers with limited experience of rocks and minerals should continue here.

Observation of the way that the minerals within the rocks have changed can provide insights into the chemical exchanges that have occurred between rocks and seawater. However, these changes can best be monitored by making *bulk chemical analyses* of representative whole-rock specimens (Table 5.1).

Basalts crystallize fairly simply from a magma and they have a comparatively narrow range of variability. However, greenschists and the other metamorphosed counterparts of basalt have much more variable chemical composition because they form by reaction with differing amounts of seawater at various temperatures for different periods of time. Hence, in Table 5.1 a meaningful average basalt analysis can be given, but greenschists have to be illustrated by a range of analyses.

QUESTION 5.2   (a)   Examine Table 5.1 and try to decide which constituents are most obviously (i) added to basalt from seawater, and (ii) leached from basalt by seawater, during hydrothermal circulation.

(b)   How might metamorphic greenschists and amphibolites, formed *within* oceanic crust by reaction with hot circulating seawater, come to be exposed at the sea-bed, either for sampling in dredge hauls or for direct examination from submersibles?

Table 5.1   Chemical analyses of oceanic basalt and greenschists (in weight %).

	Average basalt		Greenschists			
SiO$_2$	49.92	49.11	42.45	49.39	46.95	49.25
TiO$_2$	1.53	0.49	2.19	0.85	1.46	0.76
Al$_2$O$_3$	15.63	16.72	16.98	16.03	16.58	15.90
Fe$_2$O$_3$	1.65	6.82	11.84	6.80	10.26	2.67
FeO	8.19	3.69	5.69	2.42	1.63	5.83
MnO	0.17	0.17	0.19	0.17	0.05	0.16
MgO	7.65	10.96	11.73	15.42	15.13	8.71
CaO	11.17	6.09	3.18	0.29	0.27	10.44
Na$_2$O	2.75	2.57	0.80	1.34	0.11	3.83
K$_2$O	0.16	0.05	0.01	0.01	0.05	0.02
H$_2$O	0.95	4.18	7.09	7.46	8.47	3.50
	99.77	100.85	102.15	100.18	100.96	101.07

Notes

1   For this type of analysis, rock specimens are first crushed to a fine powder and then analysed for individual chemical elements.

2   Chemical analyses of rocks are conventionally expressed in terms of the weight of elements alone for minor constituents (those whose concentrations are measured in hundreds of parts per million (p.p.m.) or less), but in terms of the equivalent weight of element oxides for major constituents (those whose concentrations exceed about 0.01%).

3   Analyses rarely add up to exactly 100%, chiefly because every constituent is analysed separately, and no analytical technique is perfect.

4   All rocks are natural systems with inherent variability. No two basalts or greenschists are identical in either mineral make-up or chemical composition. Thus, between different samples of the same rock type, major constituents can vary by a few tenths of a per cent up to several per cent, depending on their relative abundance in the rock. Minor constituents (trace elements) are a good deal more variable: up to several hundred p.p.m. either way.

5.2.2   CHANGES IN SEAWATER

Recognition of the extent of hydrothermal circulation in the 1970s had a profound impact on ideas about the ways in which chemical elements are supplied to, removed from, and cycled within the oceans. The effect of hydrothermal reactions on seawater can be seen by comparing the composition of ordinary seawater with that of a hydrothermal vent solution (Table 5.2).

QUESTION 5.3   What evidence is there in Table 5.2(b) that hydrothermal solutions are (i) more acid, and (ii) chemically more reducing (less oxidizing) than seawater?

Comparison of Tables 5.1 and 5.2 shows that some of the elements that are quantitatively important in seawater are only minor or trace constituents in rocks. The notable exception is *sodium*, which is a major constituent of both.

Table 5.1 suggests that sodium may be either lost from rocks or gained by them during hydrothermal circulation. Experiments indicate that, in general, sodium is leached from rocks when the ratio between the total mass of water which has passed through the rock per unit mass of rock

Table 5.2(a) The major dissolved constituents of seawater.

Element	Concentration (p.p.m.)	Main dissolved species
chlorine (Cl)	19 500	Cl^-
sodium (Na)	10 500	Na^+
magnesium (Mg)	1 290	Mg^{2+}
sulphur (S)	905	SO_4^{2-}
calcium (Ca)	400	Ca^{2+}
potassium (K)	380	K^+
bromine (Br)	67	Br^-
carbon (C)	28	HCO_3^-
strontium (Sr)	8	Sr^{2+}
boron (B)	4	BO_3^{3-}
silicon (Si)	3	$Si(OH)_4$
fluorine (F)	1	F^-

Table 5.2(b) The composition of a typical hydrothermal vent solution at c. 350°C at 21°N on the East Pacific Rise, compared with normal seawater. All concentrations in parts per million, by weight. The pH of the hydrothermal solution for this vent is 4.0 whereas for normal seawater it is about 8.

Element	Hydrothermal solution	Seawater
Cl	17 300	19 500
Na	9 931	10 500
Mg	—	1 290
S (as SO_4^{2-})	—	905
S (as H_2S)	210	—
Ca	860	400
K	975	380
Sr	8	8
Si	600	3
Li	6	0.18
Rb	2	0.12
Ba	5–13	2×10^{-2}
Zn	7	5×10^{-3}
Fe	101	2×10^{-3}
Mn	33	1×10^{-4}

Notes

1 Only the major constituents in seawater are shown in Table 5.2(a). Most of the elements have been measured in seawater, and it is highly likely that the remainder will be detected as analytical techniques improve.

2 Expressing concentrations of elements in seawater in p.p.m. (by weight) is a convenient convention for some purposes, but you will find other conventions elsewhere in this Series, and in other literature.

3 Concentrations in Table 5.2(a) are reasonably representative, but you will find slightly different average values given elsewhere. However, the total concentration of dissolved constituents varies by only a few per cent throughout the oceans, and their relative proportions are remarkably constant.

(i.e. the water:rock ratio) is greater than about 10, but is taken up by rocks when the ratio is less than this.

The concentration of *potassium* is significantly higher in some hydrothermal solutions than in seawater, but only where temperatures are in excess of about 150°C. At lower temperatures, during sea-floor weathering, potassium is taken up by the rocks from seawater.

The concentration of *silicon* is also much higher in hydrothermal fluids than in seawater, reaching saturation in the solutions at the temperatures and pressures of the system within the crust. When temperature and pressure fall as the fluids rise towards the sea-bed, silica (SiO_2) is precipitated often in the form of the mineral quartz. Some of the *sulphate* is reduced to sulphide (H_2S, Table 5.2(b)), and the remainder is precipitated along with *calcium* to form the mineral anhydrite ($CaSO_4$) as the solutions move through the rocks. The sulphide will combine with iron and other metals to form insoluble metal sulphides. These are precipitated partly to build chimneys at the vents and partly as the suspended particles forming 'smoke' (Figure 5.1), where the hot water emerges from the sea-bed. As we shall see presently, these sulphides are also precipitated *within* the upper part of the crust where subsurface mixing occurs to produce warm water springs (see Section 5.3.1).

Table 5.2(b) shows that *magnesium* is entirely absent from this hydrothermal solution, having been removed from seawater and added to the rock (as you saw in Question 5.2(a)) to form the magnesium-rich metamorphic minerals described in Section 5.2.1. On the other hand, *iron* and *manganese* are both soluble under the acidic reducing conditions found in hydrothermal solutions, which accounts for their greatly increased concentrations. However, Fe^{2+} ions have the same size and charge as Mg^{2+} ions, and can occupy the same sites in crystal structures, so Fe^{2+} can follow Mg^{2+} into new minerals formed during hydrothermal metamorphism, rather than going into solution, and in some circumstances the rocks may become enriched in iron (*cf.* Table 5.1). This helps to explain why the iron:manganese ratio is between 50 and 100 in basalts but only about 3 in hydrothermal solutions (compare Tables 5.1 and 5.2(b)) which also happens to be the ratio in ocean-ridge sediments. Relatively more manganese than iron goes into solution in hydrothermal fluids. Under the oxidizing conditions of sea-floor weathering, both iron and manganese are insoluble, and are left as hydrous oxide residues ('rust') coating exposed surfaces.

The *chloride* concentration of the hydrothermal solution is significantly less than in seawater. This is consistent with the results of laboratory experiments, but there is little information about where the chloride has gone or about the mechanisms of its depletion during hydrothermal activity.

5.3 'BLACK SMOKERS'—AN EXERCISE IN PREDICTION

One of the more remarkable aspects of the oceanic hydrothermal story is that both the temperature and a close approximation to the composition of high-temperature vent solutions (Table 5.2(b)) were predicted two years before such solutions were sampled.

The first hydrothermal springs to be found in the oceans were located during 1977 in 2500m of water on the Galapagos spreading ridge in the eastern equatorial Pacific at 86°W. These were warm-water vents, where water emerges gently at exit temperatures in the range 6–20°C above the ambient bottom water temperature (2°C). Analysis of the solutions and comparison with experimental results strongly suggested that these comparatively low-temperature warm springs resulted from simple mixing between a high-temperature hydrothermal solution and the ordinary seawater which saturates the upper parts of layer 2A.

Laboratory experiments had demonstrated that magnesium is completely removed from seawater during high-temperature seawater–rock reactions, therefore any magnesium in low-temperature springs was likely to be due to mixing of magnesium-free hot hydrothermal water with ordinary seawater. As expected on this model, a negative correlation was found between temperature and magnesium concentrations of Mg^{2+} in samples from low-temperature vents. Extrapolation of this trend to zero Mg^{2+} gave an intercept on the temperature axis at 350°C, indicating that this was the temperature of the hot hydrothermal component. Similar extrapolations for other constituents allowed predictions to be made about the composition of the high-temperature solutions. The search was then on to find places where this hot component was vented without previous dilution.

Success came when 'black smokers' were first seen in 1979, beneath 2500m of water on the crest of the East Pacific Rise at 21°N. Here, solutions at about 350°C emerge from the sea-floor at up to several metres per second, through chimneys up to 10m high, composed mainly of sulphides of iron, copper and zinc. These are formed by precipitation from solution as the hydrothermal fluid mixes with the surrounding seawater. However, most of the precipitation occurs as very fine-grained sulphide particles, which form the dense black 'smoke' (Figure 5.1). Chemical analysis confirmed the compositional characteristics that had been predicted from the low-temperature Galapagos vent waters (Table 5.2(b)).

5.3.1 'BLACK SMOKERS', 'WHITE SMOKERS' AND WARM-WATER VENTS

We have seen that there are two extreme kinds of hydrothermal vent at ridge axes: the 'black smokers', where water emerges at temperatures of about 350°C or more, and precipitates mineral particles which build the vent chimney and form the black 'smoke', on contact with the bottom seawater (Figure 5.1); and the warm-water vents where the temperature of the emerging water is rarely more than 20°C, and often less, which are more common.

There is, in fact, a continuum of vent types between the two end-members. In between the extremes are the so-called 'white smokers', which have exit temperatures in the range 30–330°C. At these, the vented water may be clear, but it often produces white precipitates, dominated by barium sulphate ($BaSO_4$), with some iron sulphides (FeS and FeS_2) and silica (SiO_2).

Figure 5.7 illustrates the probable relationship between 'black smokers', 'white smokers' and warm-water vents, and shows how these hydrothermal vent systems may evolve with time. It suggests that the transition from warm-water vent to 'white' or 'black smoker' can happen at any stage in the overall evolution of a vent field. This results simply from precipitation of minerals, which reduces the permeability of the surrounding rock and isolates the upflow zone, so that the rising hot water cannot mix with cold water near the surface. The minerals precipitated within the rock include varieties of silica (SiO_2), anhydrite ($CaSO_4$), barite ($BaSO_4$), calcite ($CaCO_3$) and sulphides of iron, copper and zinc. They accumulate to form the seal around a conduit that eventually reaches the sea-bed and builds the chimney that characterizes 'black smokers'. Once the upflow is isolated, the vent waters cannot mix and so emerge at high temperatures and precipitate their dissolved elements on contact with seawater; hence the black 'smoke' of sulphide particles.

It follows that beneath hydrothermal vents there are accumulations of sulphide and sulphate minerals in small veins and pockets forming a *stockwork* which ramifies through the rocks of seismic layer 2, where the hot vent waters have mixed with the cold seawater saturating the upper part of the crust. A warm-water vent or 'white smoker' which persists without ever evolving into a 'black smoker' must develop a widely dispersed stockwork, which never becomes sufficiently impermeable to isolate the hot plume completely.

Where 'black smokers' occur, the stockwork has developed in such a way as to prevent mixing in the permeable layer 2. The result is that the

warm-water vent   **'black smoker'**

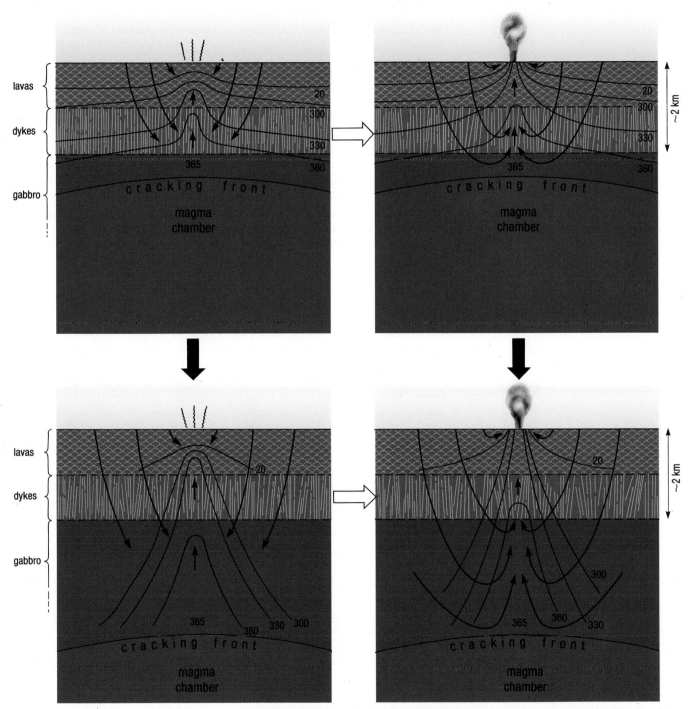

Figure 5.7   The hypothetical evolution of hydrothermal vents from warm-water vent to 'black smoker'. The depth of penetration increases with time (solid arrows), and the transition to 'black smoker' (open arrows) can occur at any time. During this transition, the vent is a 'white smoker'. A vent may stop evolving at any stage, and never develop into a 'black smoker'. Numbered lines are isotherms (in °C) of solutions circulating through the rock. The cracking front marks the maximum depth at which cracks occur and below which solutions cannot penetrate. It is approximately coincident with the roof of the magma chamber, and moves downwards as the magma in the chamber cools and the hydrothermal cell penetrates deeper.

solutions reach the surface largely unmixed, and the sulphides precipitate in the bottom seawater. Much of the material will settle on the sea-bed surrounding the vent field, but the rest will be dispersed by bottom currents over much wider areas (Section 5.6).

The oceanic crust is thus likely to be riddled with pockets of metal sulphide stockworks, at extinct vent fields. It is likely that the sulphide deposits of ophiolite complexes were deposited from hydrothermal solutions at spreading axes in ancient oceans.

A particularly striking feature of Figure 5.7 is the very sharp temperature change below the top half-kilometre or so. This is uppermost oceanic crust, where the proportion of faulted rocks, broken pillows, debris and so on is greatest. It is seismic layer 2A, with very high permeability, and there is good hydraulic continuity with the overlying seawater, which effectively saturates this layer. Within this layer (where temperatures are less than about 20°C), only sea-floor weathering reactions are widespread. Significant metamorphism does not go on at depths of less than about a kilometre below the sea-bed except in the immediate vicinity of hydrothermal conduits.

The depth of circulation of hydrothermal systems increases with time, as the rocks cool and cracks penetrate deeper into the crust, as shown by the advance of the **cracking front** in Figure 5.7. The width, spacing and rate of propagation of cracks has been a matter of much discussion and theoretical analysis, because direct observation is rather difficult. Cracks are probably not more than 1–3mm wide, with an average spacing of a few tens of cm up to 1–2m, and the rate of penetration of the cracking front has been estimated as a few metres per year. Naturally, cracks cannot propagate into a liquid magma chamber, and can extend into the gabbro of seismic layer 3 only after it has solidified.

What does the evolution of hydrothermal vents at spreading axes imply about the state of the magma chamber below?

Figure 5.7 suggests that, given sufficient time, the cracking front extends well into the gabbro layer. It shows that the top of the magma chamber can become solidified, even on the axis, due to removal of heat by the overlying hydrothermal convection system. In fact, calculations show that the rate of heat transport by hydrothermal systems *may* be sufficient to preclude the long-term persistence of liquid magma chambers, especially at slow-spreading axes. You should recall from Section 4.1.3 that no magma chamber has yet been detected by the limited number of seismic reflection profiles made across the axis of the Mid-Atlantic Ridge. Thus, in addition to being discontinuous along the axis, magma chambers may be episodic, in that hydrothermal cooling may be sufficient to crystallize completely the contents of the chamber during intervals between replenishment from below (Figure 4.21).

QUESTION 5.4 So what is the lifespan of axial hydrothermal vents? Estimates have ranged from decades to millennia, but you can make your own assessment bearing in mind that the hydrothermal solutions may penetrate to a depth of at least 5km, and the rate of advance of the cracking front has been estimated as a few metres per year. In what part of the order of magnitude scale given above does your estimate fall?

Emission rates at hydrothermal vents are in the range of $0.02–0.1 \, m s^{-1}$ for warm-water vents and $0.5–5 \, m s^{-1}$ for 'white smokers' and 'black smokers'. If the extremes of this range are taken as a rough measure of flow rates through the whole upflow zone, then it takes between about 15 minutes and 70 hours for heated seawater to rise through 5 km of crust.

Individual vent fields occupy relatively short stretches of ridge axis, probably not more than a few km at most. Vents can be very concentrated. For example, no less than 12 'black smokers' were found along an 800m stretch of the East Pacific Rise at 21°N. On the other hand, active vent fields occur at irregular intervals along ridge axes, with major gaps between them.

Why is that?

There are two reasons. First, the vent areas are much smaller than the surrounding regions through which seawater is being drawn downwards (Figure 5.5) so there have to be gaps between vent areas. Secondly, it is because the underlying magmatic activity is itself episodic and segmented, as established in Chapter 4, and the lifespan of axial magma chambers may be comparable with that of the hydrothermal systems that they 'drive'.

5.4 THE EXTENT OF HYDROTHERMAL CIRCULATION

Most 'black smoker' and 'white smoker' fields have been found along the East Pacific Rise and on the Galapagos Ridge. Fossil 'black smokers' marked by their extinct chimneys have also been located within a few km of the East Pacific Rise axis. In 1985, 'black smokers' were found for the first time on the Mid-Atlantic Ridge, near 26°N, using an unmanned instrument package which was towed only 10m above the sea-bed. This discovery followed numerous indications of hydrothermal activity, including small increases in bottom-water temperature (*cf.* Section 5.5) and strong manganese enrichment in seawater within the axial rift. The Mid-Atlantic Ridge 'black smokers' lie within the rift valley, at a depth of 3800m below sea-level, much deeper than the Pacific vents, and the sulphide deposits accumulating round the vents are significantly bigger.

From Figure 5.6 and the evidence of metamorphic rocks drilled or dredged from the ocean ridges, we know that hydrothermal activity must occur throughout the 50000km length of the spreading axis system, wherever a magma chamber develops. Over geological time the axial zone is a permanent linear heat source. Hydrothermal vent fields of the kind we have been describing occur no more than about 200m from the ridge axis, and they represent the upflow zones of Figures 5.5 and 5.7, above the hottest parts of the magma chamber. The downflow zones extend over much wider areas, covering both cooler parts of the ridge axis and more particularly the flanking regions. A few off-axis hydrothermal vents have been identified in the oceans, in association with off-axis magmatic activity at seamounts. Several hydrothermal ore deposits in ophiolite complexes appear to have formed in this way.

Newly formed crust must lie in the potential upflow zone of the axial convective system, because it is formed at the ridge axis. It moves away at a few centimetres per year.

If the spreading rate averages $2\,\mathrm{cm\,yr^{-1}}$, how far will the newly formed crust have moved from the ridge axis in 50000 years?

It will be about 1km away from the axis, and be well within the downflow zone of the convective system, being penetrated by cold unreacted seawater instead of hot reacted seawater. Newly formed crust thus starts off in the region of upflow and then moves sideways into the downflow zone.

All ocean basalt encounters both hot and cold water during its history, i.e. it passes through various stages of contact with circulating seawater, with scope for a whole range of chemical reactions and element exchanges. Some distance off-axis (perhaps a few km), convection cells may become 'locked into' the rock through which they circulate. Upflow and downflow zones remain fixed in relation to the rock, and hydrothermal circulation continues, diminishing in intensity as the crust cools with age. Much less is known about convection away from ridge axes, but detailed measurements of heat flow provide clear evidence that it occurs, as illustrated by Figure 5.8.

QUESTION 5.5 Do the peaks or troughs in the heat flow profile in Figure 5.8 represent upflow or downflow zones of convective cells?

In some places, there is evidence that off-axis circulation permeates both igneous rocks and sediments. Elsewhere, it seems that it may continue beneath a thick insulating cover of sediments, cut off from contact with the overlying seawater; the water simply goes round and round without being replenished until the system cools down enough to stop.

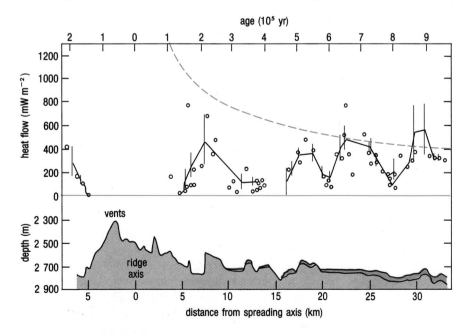

Figure 5.8 The presence of convection cells away from ridge axes can be mapped using heat-flow measurements (see Question 5.5). Open circles are individual measurements, vertical bars show the range of several measurements. The dashed line is the predicted heat flow, assuming conduction alone. The lower profile shows topography with progressively thickening sediment cover. This profile is from the Galapagos Ridge. Note the age scale is 10^5 years, so this represents a very enlarged version of the left-hand end of the graphs in Figure 5.6.

At some drill sites, when the sedimentary seal above the isolated but still-active circulation system is punctured by drilling, bottom seawater is sucked down into the hole. This must be because the circulation systems are no longer in equilibrium with the prevailing hydrostatic pressure field but are underpressured relative to their surroundings.

Profiles similar to that in Figure 5.8 have been obtained in oceanic crust as old as 55 Ma (e.g. in the Indian Ocean), which is wholly consistent with the extent of hydrothermal circulation inferred from Figure 5.6. In addition, there is evidence that the formation of new oceanic crust in back arc basins (Figure 3.8) is also associated with hydrothermal circulation (see Section 5.6). The phenomenon is therefore not confined exclusively to the main ridge systems and their flanking regions.

Fully one-third of the world's oceanic crust contains active circulation systems which are the descendants of hydrothermal systems that originated at ridge crests. With such a long history of hydrothermal activity affecting the oceanic crust, it is not surprising that the chemical changes in both rocks and seawater are extremely complex in detail and can vary a good deal from place to place.

5.4.1 VARIABILITY IN HYDROTHERMAL SYSTEMS

The basic principles of seawater convection through oceanic crust are simple, but the reality is more complex. The hot saline water of hydrothermal systems is a very powerful chemical reagent, which varies in its chemical activity through the prevailing temperature range of some 350°C and pressure range of about 500–1000 atmospheres. Moreover, the 'plumbing' of the systems can be very complicated.

Water may reach equilibrium with the rocks in one part of the system, but then react with rocks in another part, where pressure and temperature are different. Further solution and precipitation of elements can occur, including solution of elements previously precipitated, and precipitation of elements recently dissolved. The extent of interaction between water and rock is determined also by the total amount of water which has moved through the system (the water:rock ratio), and how fast it moves. Water:rock ratios can be estimated from the chemical compositions of rocks and water, but there are considerable uncertainties, because we can never be sure that the vent waters have gone through just a single reaction cycle. For example, greenschist rocks consisting almost entirely of quartz and chlorite have been recovered from the sea-bed, with SiO_2 contents of 60–70% or more, and MgO contents of 5–7% or less.

Looking back to Table 5.1, what does that imply about the water–rock reactions that formed such rocks?

The implication is that silica has been added to the rocks and magnesium has been leached from them, which is the reverse of what seems to happen normally, according to Tables 5.1 and 5.2. In spite of these complications, by using all the available information and making plausible assumptions, it has proved possible to estimate water:rock ratios in hydrothermal systems. Values range from about 1 to around 50:1 (water:rock, by mass). Because of the complex 'plumbing', it is possible for the ratio to vary even within the same system, for example if some of the water becomes isolated in a sub-cell within the main circulation.

In view of the range of water:rock ratios, we cannot expect all hot solutions emerging at 'black smoker' vents to have exactly the same composition and temperature as the one shown in Table 5.2(b). Some degree of variability is to be expected, though the relative proportions of the different constituents probably do not change much. The exit temperatures of those 'black smokers' which have been investigated lie mostly in the range 350–400°C.

Perhaps the most atypical present-day hydrothermal systems occur in the Red Sea (Figures 3.3 and 3.4), where axial deeps contain thick accumulations of metal-rich muds overlain by concentrated hydrothermal brines at temperatures of 60 °C or more and with salinities of over 300 parts per thousand, an order of magnitude greater than that of normal Red Sea water (itself rather high at nearly 40 parts per thousand). Despite their warmth, the high salinities of these brines makes them denser than normal Red Sea water, so they become ponded within the steep-sided deeps and are unable to escape. It is possible that their high salinities may in part be due to solution of salts from the nearby evaporites (Section 3.2.1).

5.4.2 HYDROTHERMAL METAMORPHISM

We have seen evidence of hydrothermal circulation in many parts of the oceans, so just what is the extent of hydrothermal metamorphism within the oceanic crust? Figure 5.7 implies that the temperature in the top 500m or so is too low for metamorphism except in the immediate vicinity of 'black smoker' conduits (remember from Section 5.2.1 that greenschist metamorphism requires temperatures of 200–400°C). However, temperatures are well within the greenschist range throughout at least the lower half of seismic layer 2, so we might expect these rocks to have been extensively metamorphosed. Unfortunately, seismic methods cannot confirm this.

QUESTION 5.6 Look at Figure 5.9 and explain why seismic methods cannot tell you whether oceanic crust has been metamorphosed or not.

So, to prove the extent of metamorphism in the oceanic crust, we need samples of the rock itself. Samples collected by dredging and submersibles are usually from the uppermost crust, subject only to sea-floor weathering. Those deriving from greater depths in the crust are brought to the sea-floor by faulting. Both metamorphosed and unmetamorphosed rocks have been recovered from regions of faulting, but they cannot be regarded as typical, because of the possibility that large faults act to concentrate the hydrothermal circulation locally. It is rare for boreholes made during the Deep Sea Drilling Project or the Ocean Drilling Project to penetrate more than a few hundred metres into the igneous oceanic crust, but one hole was drilled successfully, for 1075m into the Pacific crust 400km west of Ecuador in 1981 (DSDP hole 504B). This encountered pillow lavas and then dykes metamorphosed to greenschists in the lower part of the hole. In ophiolite complexes it is usual for metamorphism to progress from zero (sea-floor weathering only) in the upper part of the pillow lavas and to increase downwards through zeolite-bearing rocks to greenschists and sometimes amphibolites in the sheeted dykes and gabbros. This strongly suggests that the type of hydrothermal metamorphism we have been discussing is prevalent throughout much of the oceanic crust.

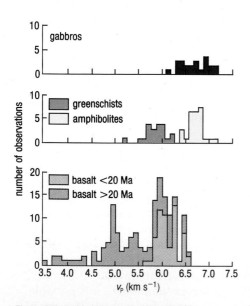

Figure 5.9 Laboratory measurements of compressional wave velocities at 1 kbar for water-saturated oceanic rocks.

5.5 MASS TRANSFER BY HYDROTHERMAL CIRCULATION

Figure 5.6 shows that most heat is lost from spreading axes and that the convective 'deficit' itself decreases exponentially away from them. The total heat deficit, i.e. the total heat lost by hydrothermal circulation in the oceans (the 'convective heat-transfer rate'), is estimated by various authorities to be between about 8×10^{19} J and 3×10^{20} J per year. It can be shown that a volume of seawater equivalent to the whole oceans must circulate through the oceanic crust in a period of the order of 10 million years to account for this.

The rate of flow of water through the crust, F (in kg yr^{-1}), necessary to achieve the required heat transfer is given by the formula:

$$F = \frac{H}{c_w(T_2 - T_1)} \tag{5.1}$$

where

H is the convective heat-transfer rate given above;

c_w is the specific heat of seawater: 4.0×10^3 J kg^{-1}°C^{-1};

T_1 is the initial temperature of seawater at the ocean floor, e.g. 2°C; and

T_2 is the temperature of seawater after circulation through the crust, e.g. 300°C

We have suggested a figure of 300°C for T_2, because (as we have seen) most of the heat is lost from the narrow zone on either side of the ridge axis. The highest temperatures are reached by the 'black smokers' on the axial rift region, but temperatures will be substantially lower on the flanks.

QUESTION 5.7 (a) With this information, and taking an approximate average value of 2×10^{20} J yr^{-1} for H, use equation 5.1 to work out the mass of seawater flowing through oceanic crust each year.

(b) There is about 1.4×10^{21} kg of water in the oceans. Show that this volume would be cycled through oceanic crust in less than 10 million years, according to your answer to part (a).

If the 1.7×10^{14} kg of seawater circulating through the oceanic crust each year (see Question 5.7) picks up the concentration of Ca^{2+} that is indicated in Table 5.2(b), then an extra 460 p.p.m. of calcium goes into solution. Thus, a total of about 7.8×10^{10} kg of calcium could be added to the oceans annually, which is not very much less than the 5×10^{11} kg of calcium introduced from rivers; although some of the hydrothermal calcium is re-precipitated in oceanic crust, as anhydrite (CaSO$_4$) and as calcite (calcium carbonate, CaCO$_3$) in cooler parts of convection cells (Sections 5.2.2, 5.3.1). Rather more sophisticated calculations of element fluxes have shown that hydrothermal activity is the major source of lithium, rubidium and manganese to the oceans, and is an important contributor of barium and silicon, as well as of calcium. It also provides the major sink for Mg^{2+} and SO$_4^{2-}$ introduced into the oceans by rivers.

Why have we not included potassium in this catalogue, as it is clearly important, according to Table 5.2(b)?

In Section 5.2.2 you read that potassium is leached from rocks by high-temperature seawater, but is added to rocks when temperatures fall below about 150 °C. Therefore, it is particularly difficult to assess its significance.

5.6 DISPERSAL OF DISSOLVED GASES AND OTHER HYDROTHERMAL EFFLUENT

Helium is a light gas that is rare in the Earth's atmosphere, because of its low atomic mass and low escape velocity. It has two stable isotopes. Helium 4 (^{4}He) is by far the commoner, being produced by radioactive decay of uranium and thorium, which are both widespread in the sediments and igneous rocks of the Earth's crust, albeit mostly in trace amounts. The much rarer helium 3 (^{3}He) was not discovered until 1938. It has two sources: one is formation by cosmic ray bombardment in the atmosphere, from which it escapes rapidly to space; the other is primordial, trapped within the Earth since its formation. However, it can escape, and ^{3}He is released by mantle outgassing in volcanic activity, including that at oceanic spreading axes.

The ratio of ^{3}He to ^{4}He is greater in hydrothermal vent waters than anywhere else. ^{3}He is the only constituent of vent solutions which is unequivocally known to have originated in the mantle and not through water–rock interactions. It thus provides a very valuable tracer for following the path of hydrothermal effluent solutions as they move away from the vents. Moreover, since helium escapes from the oceans fairly rapidly, any ^{3}He in ocean waters must have been introduced within the past few hundred thousand years, at most.

Figure 5.10 illustrates how the dispersal of the hydrothermal plume above the East Pacific Rise at 15°S can be mapped using the ^{3}He:^{4}He ratio. There is a pronounced westward flow at 2–3 km depth and the plume extends thousands of kilometres to the west.

Spreading rates are greater along this stretch of the East Pacific Rise than anywhere else in the oceans (*cf.* Chapter 3) and the ratio of ^{3}He to ^{4}He is about ten times higher than above the 'black smokers' at 21°N. Only in the Red Sea and the Gulf of California, which are young deep linear basins with restricted circulation, have higher concentrations of ^{3}He been found.

If you look back to Figure 5.4, you can see an obvious asymmetry in the distribution of metal-rich sediments about this part of the East Pacific Rise (between 5°S and 40°S). Figure 5.11 shows the independently inferred current regime *at ridge-crest depths* superimposed on this sediment pattern. The correlation seems obvious enough. However, before the ^{3}He-rich plume (Figure 5.10) was observed, circulation models for the Pacific did not include this flow regime *at mid-depth* as shown in Figure 5.11. Once independent corroboration of this circulation pattern was sought, using standard oceanographic variables (temperature, salinity, density, oxygen, nutrients) which you can read about elsewhere in this Series, it was soon established.

The plumes of water emitted from hydrothermal vents rise rapidly, because the hot solutions are very buoyant—densities of little more than $0.6 \times 10^3 \, \text{kg m}^{-3}$ have been recorded for 'black smoker' emissions.

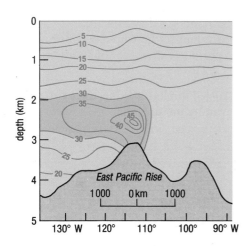

Figure 5.10 The ^{3}He-enriched plume at 15°S over the East Pacific Rise, showing the marked westward transport at mid-depths. The contours represent ^{3}He:^{4}He ratios expressed as delta-^{3}He, or δ^3He, which is a measure of the enrichment of ^{3}He in the water relative to the atmosphere.

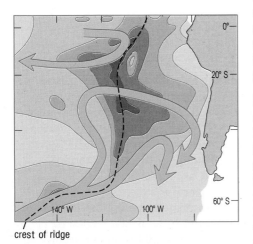

Figure 5.11 A map of the (Al+Fe+Mn)/Al ratio extracted from Figure 5.4, with independently determined current flow (arrows) at ridge–crest depths near the East Pacific Rise south of the Equator. The dashed line is the ridge crest.

However, these solutions very soon become diluted. For example, water discharged from a warm-water vent at 23°C into water at 1.8°C has a temperature of only 2.1°C (0.3°C above ambient) 10m above the vent. This represents a 70-fold dilution of the vent waters within 10m.

The ability to detect near-bottom temperature anomalies of 1°C or less was, and continues to be, a major factor in the discovery of hydrothermal vents. Sensitive instruments for measuring temperature and turbidity (due to 'smoke' particles) are used both to detect plumes and to track them away from vent fields, thus providing important information about the flow of deep currents.

We have seen that the dissipation of ^{3}He (Figure 5.10) and the distribution of metal-rich sediments (Figure 5.11) can be used as tracers for hydrothermal effluents. Other gases are also emitted, including methane (CH_4), hydrogen (H_2), carbon monoxide (CO), and nitrous oxide (N_2O).

What chemical state or condition do these four gases have in common?

They are all stable under reducing conditions and are well-known reducing agents. They may be produced either abiotically (inorganically) or by microbial activity in the vent environment.

Methane, hydrogen and carbon monoxide could be liberated by partial melting in the mantle (like ^{3}He) and they are all well-documented constituents of volcanic gases elsewhere. Alternatively, all four can be released into solution by oxidation/reduction reactions within the basalt–seawater system during hydrothermal circulation. Methane is the most abundant gas in vent waters, where $CH_4:^3$He ratios are in the order of $10^7:1$. Variations in this ratio may be due to different proportions of abiotic and microbial methane, but there may be other causes. For example, methane-enriched plumes of water found in the back-arc basin behind the Marianas island arc (Figure 1.11) are not accompanied by the marked ^{3}He enrichment that characterizes hydrothermal vent fields in the major oceans. This may mean that the upper mantle below the back-arc basin has already been depleted in ^{3}He.

In the next Chapter, you will see how information from sediment sequences can be used to reveal changes in oceanic circulation and sea-level, in response to changes in basin shape (Chapter 3) and climatic variations.

5.7 SUMMARY OF CHAPTER 5

1   Direct evidence of hydrothermal circulation through oceanic crust was not obtained until the late 1970s, but it had been predicted from several lines of indirect evidence for at least a decade before. These included the distribution of metal-rich sediments at ridge crests, metamorphic rocks dredged from ridge crests, a major conductive heat flow deficit along the whole ocean-ridge system and laboratory experiments on basalt–seawater interactions at high temperatures and sea-floor pressures.

2   All hydrothermal systems require a permeable layer of rock which allows cold water to percolate downwards over a wide area, a localized

heat source below the permeable layer, and channel-ways for the heated buoyant (low-density) plume of hot water to escape at the surface above the heat source.

3   Rock metamorphism during hydrothermal activity involves hydration, the addition of large amounts of magnesium, the loss of calcium and sometimes sodium and potassium, and also gains and losses of numerous minor and trace elements. The normal basaltic mineral assemblage (plagioclase, pyroxene, ± olivine, and basaltic glass) is transformed into various combinations of albite, chlorite, actinolite, zeolite, quartz, and other minerals, depending partly on the temperature and pressure conditions and partly on the water:rock ratios involved in the reactions. Serpentinite is formed when seawater penetrates to seismic layer 4 and hydrates the peridotite. It is plastic and of low density and is forced by pressure upwards along fractures. Eventually, it may be exposed at the sea-bed. Metamorphic rocks formed in oceanic crust are exposed along scarps of later faults and fractures.

4   During hydrothermal circulation, seawater loses all its magnesium and sulphate, and gains significant amounts of calcium and sometimes also potassium and sodium, as well as several minor elements, especially silicon, but also others (e.g. barium, rubidium, iron, and manganese). Sulphate is precipitated as calcium sulphate (anhydrite) and barium sulphate (barite) in cracks and fissures.

5   Sea-floor weathering occurs at bottom-water temperatures and involves hydration, alteration of feldspars and glass to clay minerals and (sometimes) zeolites, as well as oxidation, especially of iron and manganese to form oxide coatings.

6   Hydrothermal solutions are more acid and reducing than seawater, and carry sulphide ions. There is a continuum of types of hydrothermal vent: 'black smokers', emerging at temperatures of about 350°C or above and precipitating sulphide particles, which build vent chimneys and form black 'smoke' plumes; 'white smokers', with exit temperatures of about 30–330°C; and warm-water vents, with exit temperatures only a few degrees above normal bottom-water temperature. The latter two types are mixtures of normal seawater (which saturates the permeable upper crust), with the high-temperature solutions that elsewhere form 'black smokers'. Warm-water vents probably evolve into 'black smokers' when the upflow channel is isolated by precipitation of minerals—especially anhydrite and sulphides—and mixing with cold seawater can no longer occur.

7   Hydrothermal convection occurs throughout the ocean-ridge system and extends out to crust up to about 70 Ma old, with exponentially decreasing intensity away from ridge crests. Heat-loss calculations show that the equivalent of the whole ocean volume is cycled through oceanic crust in the space of a few million years, and the resulting fluxes of some elements into and out of the crust are comparable with (or may exceed) those of river transport to the oceans.

8   ^{3}He is a ubiquitous constituent of hydrothermal fluids, being released from the primordial 'store' within the Earth, when the upper mantle partially melts below spreading axes. It is an ideal tracer, for it has no other sources in the oceans, and can be used to trace the movement of hydrothermal effluents and hence to help provide information about

current patterns within the oceans. Other gases in hydrothermal plumes are methane (the most abundant dissolved gas), hydrogen, carbon monoxide and nitrous oxide. They may be of abiotic (inorganic) or microbial origin.

Now try the following questions to consolidate your understanding of this Chapter.

Table 5.3 Chemical analysis of average 'weathered' ocean ridge basalt in weight %.

SiO_2	47.92
TiO_2	1.84
Al_2O_3	15.95
Fe_2O_3	4.58
FeO	6.05
MnO	0.19
MgO	6.38
CaO	10.73
Na_2O	2.91
K_2O	0.53
H_2O	2.19
	99.27%

QUESTION 5.8   Compare Table 5.3 with Table 5.1. How well does it exemplify the main changes that occur in sea-floor weathering, as outlined in Sections 5.2.1 and 5.2.2?

QUESTION 5.9   Zeolites that are formed during metamorphism of oceanic crust (as opposed to those which form by sea-floor weathering) are stable in the approximate range 100–200°C, at the prevailing pressures. Chlorite is stable in the range 300–400°C. With reference to Figure 5.7, can you explain why chlorite predominates over zeolites in metamorphic rocks of the oceanic crust?

QUESTION 5.10   Assuming, for simplicity, that most of the 1.7×10^{14} kg of water per year circulating through the oceanic crust does so at spreading axes, and is largely confined to seismic layer 2, what is the average water : rock ratio (by mass) involved in the hydrothermal reactions which occur as a result of this? *Note*: At an average full spreading rate of $4 \, \text{cm yr}^{-1}$ along 50000 km of spreading axis, the volume of seismic layer 2 (total thickness 3 km) generated each year is about $6 \, \text{km}^3$. The density of basalt is about $2800 \, \text{kg m}^{-3}$.

QUESTION 5.11   The data in Table 5.2 suggest that nearly 1300 p.p.m. of magnesium are lost from seawater to rock.

(a)   Given the average flux of seawater at spreading axes of $c.$ 10^{14} kg annually, how much magnesium is lost from seawater each year?

(b)   What proportion of the annual river flux of $1.3 \times 10^{11} \, \text{kg yr}^{-1}$ is your answer to (a), and does it support the statement that hydrothermal circulation provides a major sink for magnesium in the oceans?

QUESTION 5.12   Which of the following statements are true, and which are false?

(a)   Sea-floor weathering can only occur where hydrothermal circulation is in progress.

(b)   'Black smokers' are most active along ridges in low latitudes because the Earth's surface temperatures are highest in equatorial regions.

(c)   The lowest layers of sediment sequences deposited on oceanic crust should be the most enriched in iron and manganese.

(d)   The thermal gradient through oceanic igneous crust should in general be greater at ridge–transform intersections than in the middle parts of ridge segments.

(e)   Water within the oceanic crust near ridge crests is cold bottom water saturating all the available spaces throughout crustal layer 2.

(f)   At any one time, the $^3He:^4He$ ratio in seawater will be greater than that in the atmosphere.

CHAPTER 6	PALAEOCEANOGRAPHY AND SEA-LEVEL CHANGES

In terms of geological time, individual ocean basins are short-lived features of the planet, which are continuously changing in shape and size. The systems of currents within them are even more ephemeral and those of our present-day oceans are comparative newcomers to the global scene. For instance, there can have been no Gulf Stream before about 100 Ma ago, because the North Atlantic had not yet opened wide enough to accommodate the circulation system of which it is a part.

Information about palaeoceanography (the history of the world's oceans) comes from two main sources: the changing shape of the basins themselves, as deduced from magnetic anomalies and related data; and the sediments of the sea-bed, which retain a record of past events in the overlying waters. The sedimentary record also helps to unravel the history of sea-level changes on a variety of time-scales. This is a matter of considerable concern for the future of nations with low-lying coastlines, and is of critical importance in hydrocarbon exploration on continental margins.

6.1 THE DISTRIBUTION OF SEDIMENTS

The thickness of sediments forming seismic layer 1 of the oceanic crust increases with distance from spreading axes.

What is the obvious reason for this?

The further the crust is from the ridge, the older it is and the more time has elapsed for sediments to accumulate. Sediments are no more than a few metres thick near ridge axes and even there only in depressions in the rugged topography; whereas in abyssal plain areas, sediment thicknesses of a kilometre or more are commonplace (Figure 2.19). The continental shelf–slope–rise region may be blanketed by 10 km of sediment, or more.

In Figure 6.1, which summarizes the distribution of the principal types of sediment being deposited on the deep-sea floor at the present time, these shelf–slope–rise sediments are shown white. Sediments which settle from suspension in the open oceans are known as **pelagic sediments** and, away from the polar regions, there are three main types, as indicated in Figure 6.1:

1   Calcareous biogenic (i.e. of biological origin) sediments, dominated by the remains of skeletal hard parts of **planktonic** organisms (i.e. organisms which drift passively within surface waters) formed of calcite or aragonite (calcium carbonate, $CaCO_3$).

2   Siliceous biogenic sediments, dominated by the remains of skeletal hard parts of planktonic organisms formed of silica (SiO_2).

3   Red clays, dominated by clay minerals and with a relatively small proportion of biogenic material. The red colour is due to small amounts

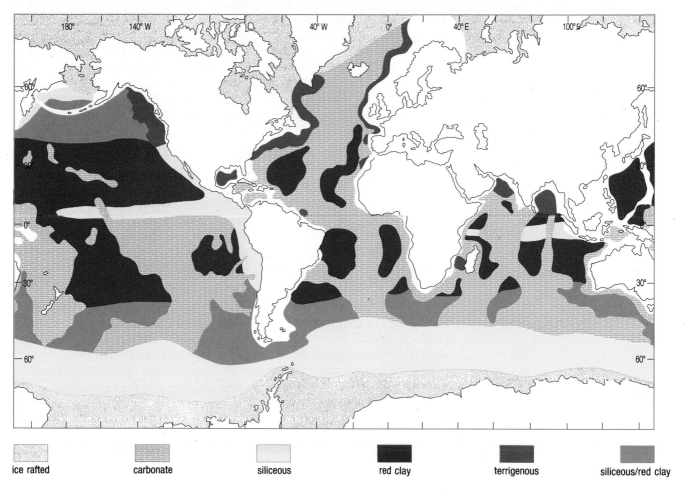

| ice rafted | carbonate | siliceous | red clay | terrigenous | siliceous/red clay |

Figure 6.1 Distribution of dominant sediment types on the floor of the present-day oceans.

of ferric iron oxide. These sediments contain the smallest particles produced by continental weathering and erosion, which are carried to the deep sea by currents or by winds. They also contain the fall-out of the fine volcanic ash from large eruptions which is carried around the globe by winds, and particles derived by the ablation of meteorites in the atmosphere. The latter accumulate at a rate equivalent to 0.1–1.0mm per million years.

The biogenic component in the red clays may be either calcareous or siliceous or both. Calcareous biogenic sediments may contain some admixture of siliceous material and vice versa, and both contain a certain amount of clay as well. Biogenic pelagic sediments used to be called oozes, which describes them rather well, but this term is falling into disuse.

The distribution of different types of sediment is principally controlled by three interrelated factors: climatic and current patterns; the distribution of nutrients and hence of organic production in surface waters; and the relative solubilities of calcite and silica, as the skeletal remains sink from the surface to the sea-bed. The dissolution of calcium carbonate is strongly depth-dependent, which accounts for its absence from deeper parts of the oceans and its relative abundance along ocean ridges. It is also temperature-dependent. Contrary to what you might expect, calcium carbonate is especially soluble in cold waters, which leads to the predominance of siliceous sediments in high latitudes.

The sediments of the deep-sea floor (Figure 6.1) are deposited on the igneous oceanic crust. Their lowest layers will almost everywhere consist of metalliferous sediments, deposited from black and white 'smokers' at ridge crests. You know from Chapter 5 that as deep-sea sediments increase in thickness with distance from ridge axes, they form the seal to hydrothermal systems. The lower parts of the sequences can be affected by the warm solutions, resulting in some alteration and recrystallization of the sediments.

You should also know from Chapter 5 that hydrothermal circulation has ceased by the time oceanic crust has reached an age of around 70 Ma. Oceanic crust at continental margins may be a great deal older than that (Figure 3.6). It therefore came as something of a surprise to marine scientists when communities of animals similar to those found at ocean-ridge crests were discovered along continental margins in the 1980s. These are supported by hydrogen sulphide and hydrocarbons produced by the anaerobic decay of organic matter within sedimentary sequences and forced out by compaction.

You read in Section 3.2 that evaporite salt deposits may form during the early stages of development of an ocean basin. These evaporites are subsequently buried by the thick sediment accumulations which form the shelf–slope–rise region. Because the salts are less dense than the sediments and because they deform plastically under pressure, they can be forced upwards as plug-like columns, doming and then punching through the sediments (Figure 6.2). These salt pillars or domes are good sites for hydrocarbon exploration, as they often provide seals for oil and gas accumulations.

Figure 6.2   Salt domes beneath eastern Louisiana (Gulf of Mexico). Some of the domes have risen through more than 10000m of sediment from salt deposits at the *base* of the sequence (note that the vertical scale is highly exaggerated). This situation is typical of that found in many continental-shelf regions.

The geological record exposed on the continents (in particular, the sediments associated with ophiolite complexes) shows that sediments similar to those in Figure 6.1 have been deposited on the floors of ocean basins since the time of the first oceans, albeit in greatly varying proportions—and of course the types of organisms making up the biogenic sediments have evolved with time.

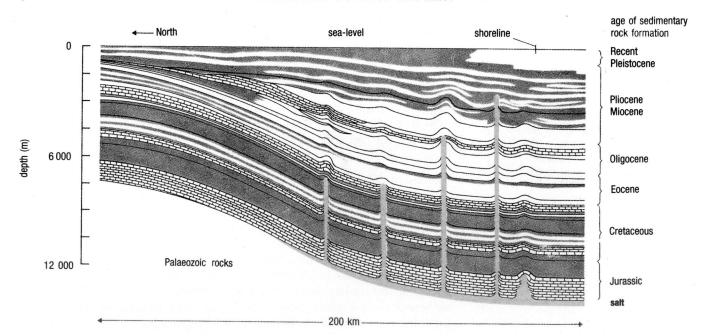

6.1.1 SEDIMENTS AND PALAEOCEANOGRAPHY

Much information about the nature and timing of events in the evolution of ocean basins can be obtained both from the sediment sequences on the sea-floor and from the remains of organisms preserved within them. By way of example, we examine a short case study in palaeoceanography: the initiation of the Antarctic Circumpolar Current.

The Antarctic Circumpolar Current is one of the major oceanic current systems, flowing continuously eastward round Antarctica and driven by the prevailing westerly winds of high southern latitudes, hence its alternative name of the West Wind Drift. The current extends to depths of 3000–4000m and owes its existence to the fact that there is deep water right round Antarctica. But it was not always so: there can have been no such current until after the southern continents had all separated from Antarctica. You can see from Figure 3.1 that the southern continents broke apart at various times after about 170 Ma ago. Figure 3.6 shows that while Australia and Antarctica were separated by ocean floor by 55Ma (if not earlier), oceanic crust appears to have begun to form between South America and Antarctica only 20Ma ago. However, there is evidence for an open circumpolar marine connection before then, based on the spread of marine fauna throughout the Antarctic Ocean. The first such evidence came from a species of the foraminiferal genus *Guembelitria*, a small planktonic organism with a maximum length of only 0.15mm (Figure 6.3). Originally restricted geographically to one part of the ancient southern ocean, its sudden appearance in sediments right the way round Antarctica provides the time mark for the opening of a free passage.

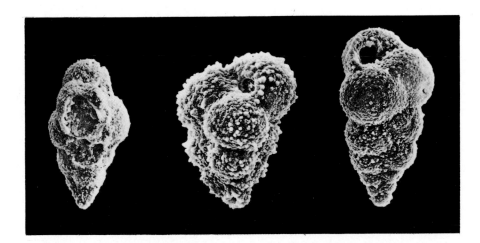

Figure 6.3   The calcium carbonate skeletal hard parts (tests) of *Guembelitria*, a planktonic foraminiferan, about 0.15mm long. Foraminiferans are single-celled organisms which may either float passively (planktonic) or live on the sea-floor (benthic).

During Lower Oligocene times (*c.* 35Ma ago), *Guembelitria* was confined to the so-called 'Austral Gulf', between Antarctica and Australia; it has been found nowhere else in the Southern Hemisphere in sediments of Lower Oligocene age. The picture changed dramatically in the Middle Oligocene, around 30Ma ago, as Figure 6.4 shows. The distribution of *Guembelitria* suggests that a circum-Antarctic current had begun to flow, although sea-floor spreading in the Drake Passage between Antarctica and South America did not begin until some 20Ma ago, according to Figure 3.6.

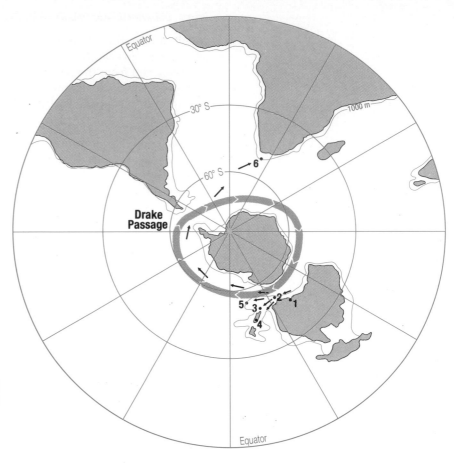

Figure 6.4   The southern oceans about 30 Ma ago, showing the spread of *Guembelitria* (red arrows) and the inauguration of the Antarctic Circumpolar Current (white arrowheads). The numbers refer to localities where *Guembelitria* is found in Middle Oligocene sediments.

Is it possible that *Guembelitria* spread not only eastwards into the Tasman Sea, as Australia and Antarctica separated, but reached southern Africa by spreading *westwards*, and so did not require the Drake Passage to be open?

It is possible, but unlikely. There is good evidence from both marine and continental sediments that the world's climatic belts and wind systems occupied much the same latitude limits then as they do today. Westerly winds probably prevailed near 60°S, and it would have been difficult for passively drifting planktonic organisms to move west against the wind-driven surface currents.

How, then, can we explain the eastward spread of *Guembelitria* round Antarctica some 10 Ma before the sea-floor age data suggest that the Drake Passage opened?

Guembelitria was a planktonic organism living in surface waters. Only a shallow channel, perhaps merely a few hundred metres deep, would have been required for it to be transported all round the continent by a wind-driven current. Such a channel could have developed by rifting without sea-floor spreading in the Drake Passage.

Evidence from the Drake Passage suggests that it did not begin to open properly until about 23 Ma ago, and deep water conditions did not develop there till as recently as 18 Ma ago; the Antarctic Circumpolar Current cannot have become fully established in its *present* configuration, which involves water depths of up to 4000 m, until then.

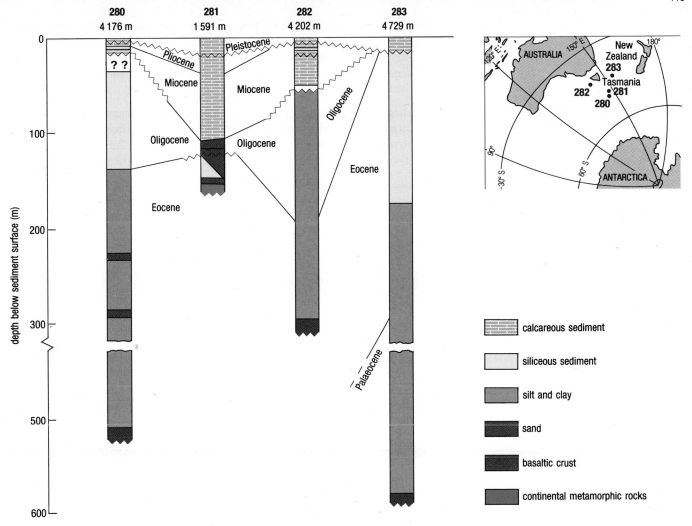

Figure 6.5 Thicknesses and ages of sediment sequences drilled at four DSDP sites near Tasmania (see inset map). Present-day water depths are given at the top of each column. The straight lines represent the age-boundaries that can be interpolated between sediment sequences; similarly the zig-zag lines mark recognizable unconformities (periods of non-deposition or erosion of sediments).

Figure 6.5 illustrates rather well how conditions changed near Tasmania, as the 'Austral Gulf' opened and water spread west into the Tasman Sea. It shows sediment sequences covering the past 40–60Ma cored from the sea-bed at the four sites indicated on the inset map. Lines between the columns correlate parts of the sequences that are of similar age, as determined mainly from studies of the fossil remains of organisms preserved in them. The zig-zag pattern indicates where intervals of time are not represented (these are called unconformities), either because sediments were not deposited or because they were deposited and then eroded away. Even at these four quite closely spaced sites, the sediment sequences show considerable differences in rates of sedimentation. Basaltic oceanic crust had formed just east of Tasmania (site 283) by the Palaeocene (c. 64Ma ago) but deep water conditions did not develop until well into the Eocene (c. 50Ma ago).

QUESTION 6.1 (a) How is it possible to deduce that last bit of information from Figure 6.5?

(b) What distinguishes site 281 from the other three sites?

(c) What sediments are accumulating in the region of the four sites today?

While data from the sediments (Figure 6.5) show that by Oligocene times (38Ma ago) the 'Austral Gulf' was deep enough and wide enough for siliceous sediments to accumulate (site 280), the evidence from the spread of *Guembelitria* suggests that the Tasman Sea connection remained closed for a further 8Ma or so.

This short case study illustrates the way in which different lines of evidence can be used by marine scientists to work out how our present-day current systems have evolved. The more detailed the information from the sediments, the more refined these reconstructions can become.

We now turn our attention to another aspect of change in the ocean basins—global and regional variations of sea-level over time, and their causes, which are an important aspect of the evolution of ocean basins.

6.2 CHANGES IN SEA-LEVEL

It is important not to confuse the *shape* of the sea-surface, as discussed in Sections 1.2.1 and 2.6, with the *level* of the sea-surface as it is perceived and measured along coastlines. Ocean bathymetry (and its effects on the geoid) changes significantly only on time-scales of 10^6–10^7 years, whereas sea-level fluctuations occur on time-scales of 10^3–10^4 years. For many purposes, therefore, we can assume the shape of the sea-surface (the geoid) to be effectively constant, merely changing in overall elevation.

Sea-level is a level of equilibrium: it is simply the level to which the ocean basins between the continental blocks are filled with seawater at any particular time. In theory, it follows that if the ocean basins are interconnected then there is a world-wide mean sea-level. A change in the volume of seawater in one ocean will affect the level in all the others. Any such world-wide change in sea-level is called a **eustatic sea-level change**.

The equilibrium level is determined by:

1   The volume of water in the oceans. This in turn is controlled by:

(a)   inputs from rainfall and snowfall, rivers, groundwater, melting ice and volcanism;

(b)   outputs through evaporation and freezing;

(c)   the temperature of the water and the amount of dissolved and suspended matter in it.

2   The shape of the container (as reflected in the hypsographic curve, Figure 2.4). This in turn is controlled by:

(a)   the global thickness and area of continental crust;

(b)   the relative thermal states (and therefore the densities) of the continental and oceanic crusts (especially the volume of active spreading ridges);

(c)   the mass of water and sediments in the oceans and the resultant load on the oceanic crust.

QUESTION 6.2 (a)   From the above lists, identify the factor that potentially has the greatest effect on the total volume of water in the oceans.

(b)   Water expands in volume by a factor of $2.1 \times 10^{-4}°C^{-1}$. Ignoring gain and loss of water by freezing and thawing, what effect would a 1°C increase in average ocean water temperature have on the total volume of water in the oceans? Is this an important factor in sea-level change?

(c)   From what you know of the age–depth relationship for oceanic crust (Figure 2.13), can you explain why, other things being equal, you might expect an ocean basin with a fast-spreading ridge to be shallower than one with a slow-spreading ridge?

6.2.1   DIFFERENT TIME-SCALES IN SEA-LEVEL CHANGES

Sea-level is subject to numerous short-term changes, some of considerable magnitude. The principal ones are tidal fluctuations, wind-generated waves, barometrically influenced surges (see below), tsunamis, freshwater floods, and even the waves produced by passing ships.

In spite of all these variations, which may amount to 10 or more metres in vertical range, it is still possible to define *mean* sea-level and to measure changes in it of the order of $1\,mm\,yr^{-1}$.

For the recent past, changes in sea-level relative to a particular stretch of shore can be determined by analysis of tide gauge records. This is a complicated task because it has to take into account numerous seasonal and occasional events as well as regular tidal fluctuations, before a reliable estimate of sea-level change can be made.

Figure 6.6 shows some of the principal variables that must be considered in arriving at a long-term analysis of sea-level change.

The fluctuations labelled 'atmospheric contributions' in Figure 6.6(c) are of the type which, on a shorter time-scale, give rise to the barometrically influenced surges mentioned above: the lower the atmospheric pressure, the higher the yearly mean sea-level. A change in atmospheric pressure of 1 mbar will change sea-level by about 1 cm (10 mm). Atmospheric pressure at sea-level can vary from about 930 mbar or less in severe cyclones (depressions) to 1080 mbar or more in strong anticyclones. The disastrous North Sea floods of 1953 resulted from a combination of a very high tide, very strong onshore winds, and very low atmospheric pressure. Yearly average atmospheric pressure differences of about 10 mbar are sufficient to account for the range of sea-level fluctuations attributed to atmospheric effects in Figure 6.6.

QUESTION 6.3   What was the average annual rise in sea-level at Esjberg over the period shown in Figure 6.6?

The trend shown in Figure 6.6 is due partly to the continued melting of glaciers and ice-caps as a result of global warming; and partly to expansion caused by the rise in average temperature of near-surface ocean waters (*cf.* Question 6.2(b)). Global temperature variations have been going on at various rates and at various scales throughout geological time. It is likely that the present rise is exacerbated by the 'greenhouse effect' of carbon dioxide and other gases released into the atmosphere by combustion of fossil fuels, and by deforestation which reduces the biosphere's capacity to remove carbon dioxide from the atmosphere. This

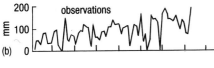

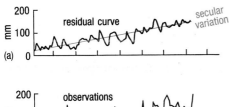

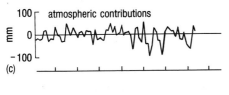

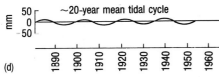

Figure 6.6   Changes in the mean sea-level at Esbjerg (Denmark), 1889–1962. (a) The progressive (secular) trend (blue line) deduced from the residual fluctuations is obtained when (b) the observed measurements are corrected for (c) changes in mean annual atmospheric pressure, and (d) long-term tidal fluctuations, due to orbital and other astronomical factors. (The significance of using a graph which ends in 1962 will become apparent later)

progressive marine inundation could eventually pose a major threat to the world's littoral populations.

It may be, however, that the *rate* of rise in sea-level is actually decreasing, because, as populations, urbanization and industrialization increase, humanity's water needs grow. An enormous volume of rain and river water is being delayed on its oceanward path by diversion into reservoirs and irrigation projects (whence some of it infiltrates into the ground). This volume, about 375 km^3 per year by the mid-1980s, is a significant addition to the total volume of freshwater in the hydrological cycle.

Figure 6.7 summarizes the results of some calculations based on known and estimated volumes of water stored and used for irrigation world-wide, and illustrates the computed effect on the rate of sea-level rise.

Figure 6.7 Suppression of global sea-level increase from 1932 to 1982 due to water storage in reservoirs and irrigation projects. Line (c) is the global sea-level rise, based on published data to 1982 and extrapolated. Line (a) is the storage capacity in large reservoirs world-wide. Line (b) is the estimated quantity of water in smaller reservoirs and irrigation projects. Line (d) is the sum of (a) + (b) + (c), to show what the sea-level rise would be if this storage had not taken place.

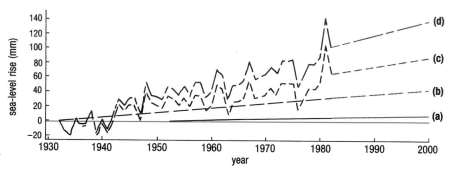

QUESTION 6.4 (a) Estimate the rates of sea-level rise for the period 1932–1982 indicated by lines (c) and (d) on Figure 6.7, and compare them with your answer to Question 6.3.

(b) What are the implications of your answer to (a)?

Let us turn now from the future prospects to consider past sea-level changes. The present gradual rise is merely part of a trend that has been going on for about the past 20000 years. In the longer perspective, sea-level has been rising and falling continuously on time-scales varying from thousands to millions of years.

6.2.2 THE POST-GLACIAL RISE IN SEA-LEVEL

There have been several times in Earth history when ice-sheets were widespread in high latitudes. The most recent of these 'ice ages' was during the Quaternary (during the past 2Ma), and it may not be over yet. Within this period there have been several individual **glaciations**, i.e. advances of the polar ice-caps, followed by their retreat. The most recent such glacial episode lasted from about 120000 until about 20000 years ago. The effects on sea-level of the retreat of this ice have been documented in detail in many parts of the world.

The initial rise in sea-level, and presumably the initial melting of the continental ice, seems to have been relatively rapid from 18000 years ago, gradually slowing to around 0.2 mm yr^{-1} about 6000 years ago, but continuing to rise slowly up to the present day (Figure 6.8). Figures 6.6 and 6.9 illustrate the modern rise for the southern parts of the North Sea. In addition to the increased volume of water due to melted ice, part of the continued rise is due to the long-term subsidence of the southern North Sea basin, which has led to the accumulation of 1000m of sediment in the past million years or so. However, local factors complicate the record of post-glacial changes in sea-level around the southern North Sea and make it difficult to obtain a totally independent frame of reference against which sea-level changes may be measured.

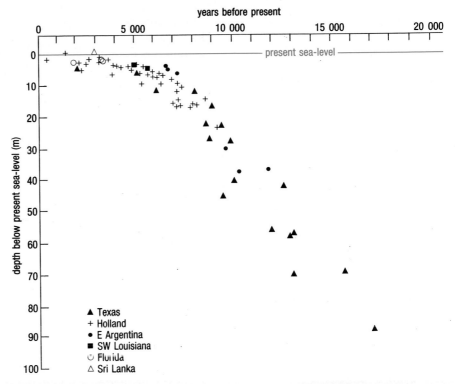

Figure 6.8   The chronology of the sea-level over the past 18000 years deduced from ^{14}C dating of peat and the shells of shallow-water marine organisms. All locations are in stable areas of the Earth's crust, where major Earth movements have been negligible, but there is still a wide scatter in the data, which define a *trend*, rather than a single global curve for sea-level.

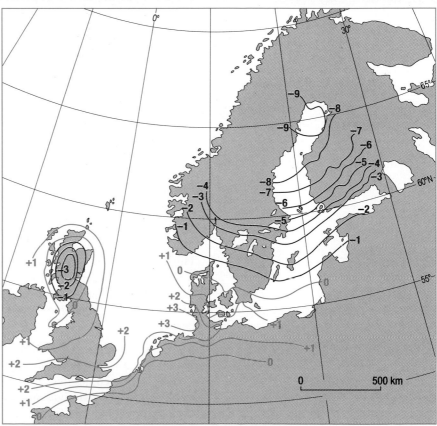

Figure 6.9   The present annual rate of change in sea-level in north-western Europe. The contours represent rises (positive) or falls (negative) in mm per year.

124

QUESTION 6.5 From Figure 6.9, where is sea-level still rising in north-western Europe? Can you explain why it is actually falling around most of Scandinavia?

The additional loading on the oceanic crust and continental shelves when melt waters return to the oceans further complicates the situation at continental margins. Even in the middle of the ocean it appears that sea-level rises more slowly (relative to an island) than would be expected from the rate of return of water to the oceans. This is because the additional weight of returning water can isostatically depress the ocean crust.

There are, in fact, no simple answers as to how local isostatic adjustments combine with eustatic changes (i.e. changes in sea-level that can be identified around the world) in the record of post-glacial changes in sea-level.

6.2.3 MEASURING QUATERNARY CHANGES IN SEA-LEVEL

Further back in the Quaternary, data are insufficient to assess the contribution of isostatic adjustments to changes in sea-level. So, in a sense, the elucidation of sea-level changes becomes easier: we have only the evidence of 'eustatic' (ocean-wide average) changes to work with. The principal feature of the Quaternary period has been the climatic fluctuations responsible for the glaciations of the past two million years.

A method originally devised in the 1960s, in order to estimate past marine temperatures, has proved to be of considerable value also in the study of sea-level fluctuations. It depends on the differential incorporation of the ^{18}O and ^{16}O isotopes of oxygen into the calcium carbonate ($CaCO_3$) which many marine organisms use to form their shelly (skeletal) hard parts.

Oxygen isotopes and the ^{18}O:^{16}O ratio: Oxygen has three stable isotopes with atomic mass numbers of 16, 17 and 18 respectively. 99.763% of natural oxygen is made up of ^{16}O, and ^{18}O makes up most of the balance at 0.204%. Oxygen-isotope studies rely on small differences in the ^{18}O:^{16}O ratio in different samples. Although these differences are minute, they can be measured very precisely using a mass spectrometer.

We shall look at the quantitative aspects presently, but we concentrate first on the natural processes which control variations in the oxygen-isotope ratio. Marine organisms which form skeletal hard parts of calcium carbonate incorporate different proportions of the ^{18}O and ^{16}O from the water according to its temperature: the lower the temperature, the greater the ^{18}O:^{16}O ratio in the calcium carbonate of the skeletal material.

Here, then, is a potential method of measuring past temperature of ocean waters, especially if organisms with a wide distribution can be used. Planktonic foraminiferans are ideal, because they are abundant, widespread, and have hard parts of calcium carbonate, generally of the calcite variety.

By measuring the oxygen-isotope composition of the skeletal calcite from planktonic foraminiferans from Quaternary sediments on the ocean floors (where they accumulated upon sinking after death), it should be possible to obtain a record of fluctuations in surface-water temperatures. It should then be possible to interpret this record in terms of glacial maxima and minima, and hence in terms of eustatic changes in sea-level (Figure 6.10). The picture proves to be not quite as simple as this, however, and there is a refinement that enables the interpretations to be put on a different basis.

Initially, it was thought that most of the variation in the $^{18}O{:}^{16}O$ ratio for planktonic foraminiferans could be attributed simply to the temperature of the surface waters where they lived. But when the skeletons of **benthic** (bottom-dwelling) foraminiferans in Quaternary sediments were examined, these also showed almost as great a variation in oxygen-isotope composition. This was a problem, because it seemed to indicate a greater range than is likely in the temperatures of the oceanic bottom waters.

The temperature of present-day oceanic bottom waters nearly everywhere lies in the range of 0–2°C. If (as seems likely) it did not change significantly between glacial and interglacial periods within the Quaternary, how can the similarity between the variation in the oxygen-isotope ratios in calcite from both benthic and planktonic organisms be explained?

To answer that question we begin by recalling that oxygen makes up about 90% of water by weight, and that ^{16}O is lighter than ^{18}O. Accordingly, water vapour tends to be enriched in molecules containing the *lighter* oxygen isotope ($H_2^{16}O$) and depleted in molecules with the heavier isotope ($H_2^{18}O$) relative to the liquid from which it evaporated (i.e. the oceans). When water vapour condenses to produce precipitation, there is a slight degree of fractionation in the opposite sense, i.e. the water condensing from vapour and then precipitating is slightly enriched in molecules with the *heavier* isotope ($H_2^{18}O$) relative to the remaining vapour, but because of the different temperatures at which the evaporation and condensation occur the amount of isotopic fractionation is greater during evaporation. Thus, the atmospheric evaporation–precipitation cycle results in net fractionation of oxygen isotopes and precipitated water is richer in $H_2^{16}O$ than the seawater from which it evaporated.

On a global scale, when ^{16}O-enriched water vapour is precipitated as snow and builds up to form glaciers and ice-caps, then the ice will be relatively depleted in ^{18}O (low $^{18}O{:}^{16}O$ ratio), while the oceans will be relatively enriched in ^{18}O (high $^{18}O{:}^{16}O$ ratio). The larger the ice-caps, the larger the proportion of ^{16}O removed from seawater (as $H_2^{16}O$), and the more the $^{18}O{:}^{16}O$ ratio of the seawater will increase.

Thus, the *range* of variation of oxygen-isotope ratios in Quaternary planktonic and benthic foraminiferans is similar, because the ratios are

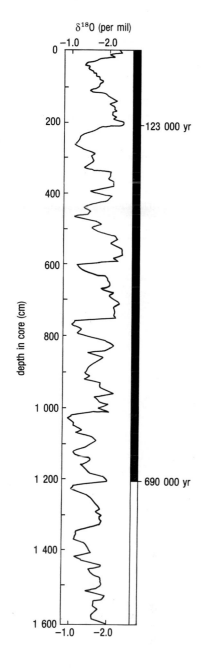

Figure 6.10 The oxygen-isotope composition (relative to a standard) of the planktonic foraminiferan *Globigerinoides sacculifera*, determined in a complete 1600 cm core which spans nearly the last million years. Isotopic composition is expressed not as a simple ratio, but as $\delta^{18}O$, which is explained in the text. Plots based on benthic (bottom-dwelling) foraminiferans during the same period look almost identical. The peak at 123 000 years ago represents the last interglacial sea-level maximum, and the minimum about 50 cm from the top of the core represents the last sea-level minimum before the present, during the last glaciation. The black band represents the most recent interval of positive polarity in the Earth's field (the Brunhes epoch) and the white band underneath is the preceding negative polarity interval (the Matuyama epoch). Correlation with the magnetic polarity time-scale has helped in the accurate dating of the sedimentation of this and other sediment cores.

reflecting changes in the volume of water 'locked up' in ice-caps and glaciers, rather than direct temperature effects. However, there is an implicit temperature effect in this relationship, because the lower the global surface temperature, the more ice we would expect to be present in ice-caps and glaciers.

We must not forget the organic effect with which we started. All organisms secreting calcium carbonate have higher $^{18}O{:}^{16}O$ ratios in cold than in warm water. However, this ratio is different for different species, even in water of the same temperature: i.e. two different species in the same body of water have different $^{18}O{:}^{16}O$ ratios: but for both species the ratios are always greater in cold than in warm water.

All this may seem complicated, but the principles are simple enough, and we can summarize the main points as follows:

1   The larger the ice-caps, the greater the $^{18}O{:}^{16}O$ ratio of seawater.

2   The higher the $^{18}O{:}^{16}O$ ratio of seawater, the higher it will be in the skeletons of marine organisms secreting calcium carbonate shells, and vice versa.

3   The lower the temperature of the water, the higher will be the $^{18}O{:}^{16}O$ ratio in the carbonate skeleton of a given species. However, the ratios will not necessarily be identical in different species inhabiting water of the same temperature.

The conventional measurement, $\delta^{18}O$: To achieve uniformity of mass spectrometer results between laboratories, the measured $^{18}O{:}^{16}O$ ratios must be calibrated against a standard sample. This can be present-day Standard Mean Ocean Water (SMOW), but for work relating to sea-level change, the so-called PDB standard is often used (B stands for a belemnite—a cephalopod mollusc fossil—from the Cretaceous Peedee Formation in the USA).

Isotopic ratios are conventionally reported as δ (delta) values, which are expressed in parts per thousand (or per mil) terms rather than per cent, as follows:

$$\delta^{18}O = \frac{(^{18}O/^{16}O)\ \text{sample} - (^{18}O/^{16}O)\ \text{standard}}{(^{18}O/^{16}O)\ \text{standard}} \times 1000 \qquad (6.1)$$

QUESTION 6.6   (a)   If the standard is SMOW, does a positive $\delta^{18}O$ value indicate enrichment or depletion in ^{18}O relative to standard seawater?

(b)   What does a negative $\delta^{18}O$ (relative to SMOW) value tell us?

(c)   Would you expect the calcareous skeletons of organisms living at the present day in cold water to have higher or lower $\delta^{18}O$ values than those of the same species living in warm water?

(d)   Would you expect the $\delta^{18}O$ value of polar ice to be positive or negative relative to SMOW?

We can expand on your answer to part (d): the lower the temperature at which evaporation occurs, the greater the enrichment of ^{16}O in water vapour. Rain in tropical regions gives $\delta^{18}O$ values close to zero (i.e. it is isotopically very similar to SMOW), while polar snow and ice can have values ranging from -30 per mil (the Greenland ice-cap) to -50 per mil (the South Pole).

Applications of the δ¹⁸O values: To recap briefly, we can say that most of the observed variation of $\delta^{18}O$ in foraminiferal skeletons during the Quaternary must be due to the enhanced differential incorporation of ^{16}O into the ice of the polar ice-caps during glacial periods, leaving the oceans relatively enriched in the heavy isotope, ^{18}O. According to this interpretation, the $^{18}O:^{16}O$ ratio of foraminiferans, and especially that of benthic species which live in low-temperature bottom water, can be taken as a measure of the amount of ocean water held in ice-sheets at any given time, and hence as an indicator of global sea-level.

Comparison of the oxygen-isotope composition of foraminiferans at the peak of the last glaciation (when the sea-level was at a minimum) with the composition of modern foraminiferans allows a direct relationship to be established between isotopic composition and sea-level. A difference in $\delta^{18}O$ of 0.1 per mil is found to be equivalent to a 10m change in sea-level, and this relationship can be used to estimate global variations in sea-level resulting from changes in polar ice volumes for the past two million years.

Although benthic foraminiferans are more reliable for this purpose, useful information can also be obtained from planktonic foraminiferans, which are more abundant (Figure 6.10)

QUESTION 6.7 (a) Why are the $\delta^{18}O$ values of benthic foraminiferans considered to be more useful than those of planktonic foraminiferans in determining past changes in sea-level?

(b) Using Figure 6.10, estimate the extent of the sea-level change for (i) the last post-glacial rise in sea-level, and (ii) the preceding major rise in sea-level at about 123 000 years ago associated with the interglacial warm period.

(c) Looking at Figure 6.10, would you say that among the numerous rises and falls of sea-level in the past 800 000 years (i) the last major fluctuation was especially noteworthy and (ii) the rises are more or less abrupt than the falls?

The Quaternary fluctuations in sea-level of 100m or more (Figure 6.10) were the result of about $50 \times 10^6 km^3$ of water being alternately withdrawn from and returned to the oceans. If the remaining polar ice-caps were to melt, the $30 \times 10^6 km^3$ or so of water locked up in them would raise the world's sea-level by about a further 60m. There is good evidence that the short-term fluctuations characterizing ice ages are superimposed on longer-term steady growth and decay of the polar ice-caps. We can demonstrate this by summarizing the history of the Antarctic ice-sheet.

6.2.4   THE GROWTH OF AN ICE-SHEET: ANTARCTICA

Evidence for the rate of growth of the Antarctic ice-sheet comes mainly from the nature of the sediments on the sea-floor around Antarctica (Figure 6.1), supported by oxygen-isotope analyses of foraminiferal remains in those sediments. The stratigraphic distribution of ice-transported sediments around Antarctica has been well-documented. These sediments are characterized by:

1   The presence of exotic rock fragments, which can only have been brought from the Antarctic continent by ice which then melted and deposited its sedimentary load.

2   Poor sorting (i.e. having a wide range of particle sizes), quite unlike those found along continental margins in non-glaciated areas.

3   Quartz grains with distinctive surface features that are typical of glaciated regions (and can be identified with a scanning electron microscope).

Reports of ice-transported debris in sediments of late Eocene age (*c.* 40 Ma ago) off western Antarctica suggest that this region may have been partly glaciated at that time, although the oxygen-isotope data indicate that there were no large accumulations of ice. The first definite

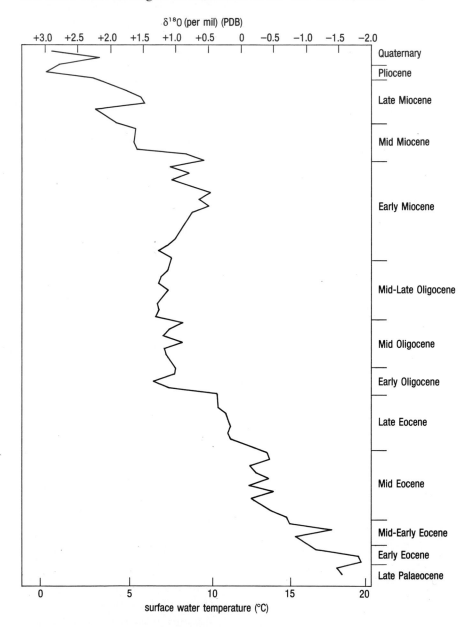

Figure 6.11   Oxygen-isotope composition and surface-water temperatures for the Antarctic region, obtained from planktonic foraminiferal remains in DSDP cores. Arbitrary samples are plotted at constant intervals, so the spacing of periods along the side is not proportional to thickness of sediment or to age differences (see Appendix for geological ages in Ma). *Note*: The standard used for the isotope ratios is different from that used in Figure 6.10 which covers a much shorter time-scale and lies entirely within the Pleistocene; but the pattern would be similar whichever standard were used.

identification of debris transported by ice is in late Oligocene sediments (*c.* 25 Ma ago) off eastern Antarctica. The proportion of glacially derived material increases progressively in younger sediments and becomes more widespread, reaching its most extensive distribution in the Quaternary. Thus, the Antarctic ice-sheet was in existence at least 25 Ma ago, and probably earlier. The sedimentary evidence for the growth of the ice-sheet is supported by the oxygen-isotope data, which indicate a progressive decrease in the temperature of southern ocean waters.

QUESTION 6.8   (a)   According to Figure 6.11, what was the surface-water temperature when the first definite signs of ice-transported debris appeared in the sedimentary record?

(b)   When would you say that the Antarctic ice-sheet began to grow rapidly towards its maximum development in the Pleistocene ice age?

It is possible that by the late Oligocene (25 Ma ago), the Antarctic ice-sheet was big enough to have a noticeable effect on sea-level, and it has been suggested that some of the unconformities in the sedimentary record (e.g. Figure 6.5) after then—especially in sequences of sediments on continental shelves—may be a result of this.

As you may have deduced in attempting Question 6.8, the rate of cooling increased considerably in the mid-Miocene, and the Antarctic ice-sheet began to grow more rapidly. This development has been linked to the final separation of South America from Antarctica, initiation of the Antarctic Circumpolar Current (Section 6.1.1) and thermal isolation of the Antarctic continent from warmer waters in the north. In the Northern Hemisphere at about the same time, glacial conditions became more widespread in the mountainous regions, although the great continental ice-sheets did not appear there until much later—about 3 Ma ago—by which time the Antarctic ice-sheet was already much the same size as it is today.

Even the rather coarse scale of Figure 6.11 suggests that the growth of the ice-sheet was not uniform. Indeed, the alternation of warm interglacials and cold glacials indicated by Figure 6.10 extends further back in time. About 5 Ma ago, those fluctuations were associated with one of the most remarkable effects of sea-level change to be found in the geological record.

6.2.5   THE SALINITY CRISIS IN THE MEDITERRANEAN

Early in the Miocene (*c.* 20 Ma ago), the Arabian plate impinged on the Eurasian plate, cutting off the Mediterranean from the Tethys and Pacific Oceans to the east (Figure 3.1). The Mediterranean became a landlocked sea, its only connection with open water being the shallow straits to the west (themselves tending to close as Africa moved northwards relative to Europe), much as today. The loss of the eastern marine connection led to the development of somewhat drier climatic conditions throughout the Mediterranean region. Evidence for this comes in part from the thick sequence of Miocene evaporites in the Red Sea basin, which was connected to the Mediterranean during most of the Miocene (Section 3.2.1). Towards the end of the Miocene (*c.* 5–6 Ma ago), this connection was broken and a passage opened up between the Red Sea and the Indian Ocean to the south. Around this time, deposition of evaporites ceased in the Red Sea but commenced in the Mediterranean: great thicknesses were deposited over wide areas of the Mediterranean basin over a period of about a million years. This event involved the virtually complete drying up of the Mediterranean and the deposition of evaporites under extremely saline conditions—hence the term 'salinity crisis'.

It is difficult to imagine that the Mediterranean could ever have dried up completely, especially as it is now more than 3km deep in places, with an average depth of about 1.5km.

Consider how long the Mediterranean might take to dry up completely, if it were to become isolated from the Atlantic Ocean again.

The surface area of the Mediterranean is about $2.5 \times 10^6 km^2$, and its average depth of 1.5km gives a total volume of water of about $3.75 \times 10^6 km^3$. Evaporation is considerably in excess of precipitation at these latitudes: the annual loss of water from the Mediterranean by evaporation (E) has been worked out at $4.7 \times 10^3 km^3$, and the annual precipitation (P) is $1.2 \times 10^3 km^3$. The difference between the two is:

$$E - P = (4.7 \times 10^3) - (1.2 \times 10^3)$$
$$= 3.5 \times 10^3 km^3 yr^{-1}$$

About $0.25 \times 10^3 km3 yr^{-1}$ of water is added to the Mediterranean by rivers and from the Black Sea; this reduces the net loss to about $3.25 \times 10^3 km^3 yr^{-1}$. At the present time, of course, this net loss of water by evaporation is made good by water flowing in through the Straits of Gibraltar.

QUESTION 6.9   Suppose the Straits of Gibraltar were to be closed, cutting off this flow from the Atlantic. How long would it take for the Mediterranean to dry up?

Your answer shows that major changes affecting the Earth's surface do not always require a great time—they can sometimes happen over the span of human dynasties. There is no question that the Mediterranean *could* completely dry up in a millennium or two. The evidence that it did become dry comes mainly from buried river gorges a kilometre or so below the valleys of large present-day rivers such as the Nile and Rhône and from the late-Miocene evaporite salt deposits, thicker than 1km, within the sediment sequences sampled by drilling over much of the Mediterranean. In many places, they form salt domes (Figure 6.12).

We now encounter something of a problem. There is no way that the evaporation of $3.75 \times 10^6 km^3$ of seawater (the present volume of water in the basin) can produce a 1km-thick layer of salts on the Mediterranean floor.

How can we be so sure?

There is some 35g of dissolved solids in every litre of normal seawater, and there are 10^{12} litres in a cubic kilometre. So, the amount of salt in the seawater in the Mediterranean today is approximately:

$$35 \times 10^{12} \times 3.75 \times 10^6 \approx 130 \times 10^{18} g$$
$$\text{or } 1.3 \times 10^{17} \text{ kg}$$

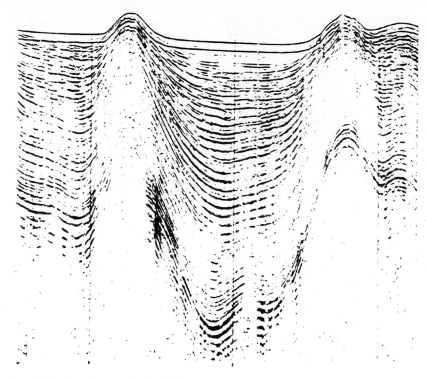

Figure 6.12   A continuous seismic reflection profile of part of the eastern Mediterranean, showing salt domes (*cf.* Figure 6.2). Some protrude as knolls above the sea-bed; others are still deeply buried. The profile is about 30 km long and the vertical exaggeration is ×6.

and it is unlikely to have been very different around the end of the Miocene.

A reasonable average density for such salts is $2 \times 10^3 \text{kg m}^{-3}$, so that amount of salts would have a volume of:

$$\frac{1.3 \times 10^{17}}{2 \times 10^3} = 6.5 \times 10^{13} \text{m}^3 = 6.5 \times 10^4 \text{km}^3$$

QUESTION 6.10   Assume that these evaporites accumulated over four-fifths of the Mediterranean floor, say an area of $2 \times 10^6 \text{km}^2$. What thickness would that volume of evaporites attain?

The answer to Question 6.10 suggests that the evaporation of something like 30–35 times the volume of water in the present Mediterranean would be required to produce the 1 km-thickness of evaporites found in the sediment sequences.

Thus, the Mediterranean cannot have been totally cut off from the Atlantic throughout the time of the 'salinity crisis' between about 5 and 6 Ma ago. There must have been some influx of Atlantic water to account for the amount of salts deposited, perhaps even enough to re-establish normal marine conditions for short periods. This is quite possible, considering that once the connection was broken again, the basin could dry up in little more than a thousand years. Huge volumes of seawater must have been supplied intermittently in order to enable the total thickness of evaporites to accumulate. The basin must have been very rapidly filled whenever the Atlantic connection was re-established; the

Figure 6.13   The Gibraltar Waterfall. (*Artist:* Guy Billout.)

Gibraltar 'waterfall' is thought to have been about a hundred times larger than the Victoria Falls on the Zambesi river (Figure 6.13). The average rate of supply of seawater, however, was below the average rate of evaporation throughout the million years when the evaporites were being deposited.

All this changed abruptly about 5 Ma ago, when the supply began to exceed the amount removed through evaporation and the whole basin was filled up again; normal marine conditions were restored and muds and deep-water carbonate sediments were deposited once more. The Atlantic connection became deep enough for cold deep water to gain access to the Mediterranean throughout most of the Pliocene, but about a million years ago the Gibraltar sill was uplifted and this deep supply ceased. Since that time, the Mediterranean has gradually acquired its present characteristics.

Evaporites, $\delta^{18}O$ changes, ice volume and sea-level: Oxygen-isotope data have been used to relate changes in the volume of glaciers and polar ice-caps to the changing sea-levels which were responsible for bringing about what is now widely known as the Messinian salinity crisis (because it occurred during a part of the late Miocene known as the Messinian).

The evaporites were deposited in two discrete periods. The sedimentological evidence from within the sequence provides abundant evidence of repeated phases of flooding, desiccation, and subaerial erosion within each period.

Figure 6.14   Oxygen-isotope data from the benthic foraminiferans *Planulina wuellerstorfi* and/or *Cibicdoides kullenbergi* from Site 588, DSDP Leg 90, south-west Pacific. The sampling interval between depths of 80 and 160 m is about 20000–25000 years. The raw $\delta^{18}O$ values clearly show that the late Miocene–early Pliocene was characterized by high-frequency variation in $\delta^{18}O$ values. The 3- and 5-point running averages of the data reveal the general trends in $\delta^{18}O$ values, and resemble glacial–interglacial cycles. The two main periods of evaporite deposition in the Mediterranean are shown for comparison.

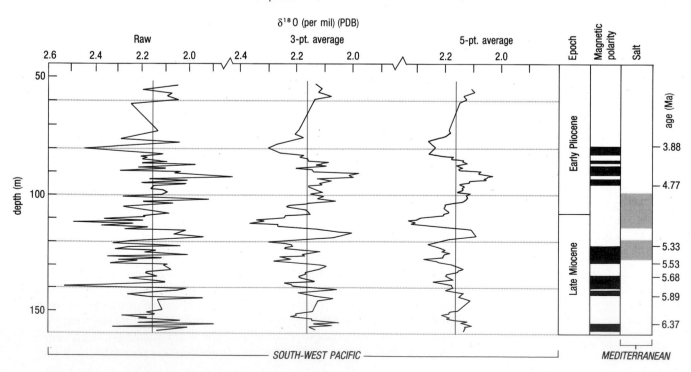

Towards the end of the deposition of the lower evaporites, the supply of water from the Atlantic was probably cut off completely. The resulting desiccation is represented by a widespread unconformity (an interval of non-deposition and erosion). This was in turn followed by a marine transgression (influx of the sea), when the Mediterranean may have been refilled to the brim and deep-water sediments accumulated for a while. There was then another period of repeated desiccation when the upper evaporites were deposited. Evaporite deposition finally ceased in the early Pliocene, and seawater refilled the Mediterranean, restoring the conditions for pelagic (deep-water) sedimentation.

Evaporite formation was initiated in the Mediterranean when it became isolated from the world ocean, by a combination of long-term tectonic uplift in the west (associated with the northward movement of the African plate), and a global fall in sea-level. The successive inundations and desiccations during the rest of the Miocene and the earliest Pliocene were controlled by fluctuating sea-levels due to changes in global ice volume. These changing sea-levels are revealed by oxygen-isotope measurements on samples of benthic foraminiferans from sediments with very accurately known ages, from both the Atlantic and Pacific Oceans. Some of the results and the correlation with the Messinian evaporites are shown in Figure 6.14.

QUESTION 6.11   (a)   Looking especially at the right-hand curve of oxygen-isotope data in Figure 6.14, why do you think that $\delta^{18}O$ values were high during the two periods of evaporite deposition but low before, after and in between them?

(b)   How does the left-hand curve of oxygen-isotope data help to explain the thickness of the evaporite sequences?

(c)   Approximately what is the time-scale of the cycles in the left-hand curve? (Note that the age-scale on Figure 6.14 is not uniform.) How does it compare with that in Figure 6.10?

(d)   What was the approximate duration of the Messinian salinity crisis, according to Figure 6.14?

(e)   Figure 6.14 indicates a major fall in sea-level at around 4 Ma ago. Suggest why this did not lead to yet another phase of desiccation and evaporite deposition.

(f)   From your earlier reading in this Chapter, where did most of the ice probably accumulate to bring about the falls in sea-level?

Now we will consider briefly whether the uppermost part of the Miocene evaporite sequence in the Red Sea (Section 3.2.1) was deposited in a similar dried-out basin. This seems very unlikely for a variety of reasons. The Red Sea evaporites were almost certainly deposited in a shallow, subsiding, basin, but the sea-level throughout the Miocene was very close to what it is now. There are coral reefs along the margins of the Red Sea which provide evidence of that. The Red Sea evaporites must have reached practically their present thickness towards the end of the Miocene (5–6 Ma ago) and so they pre-date the onset of major global sea-level falls due to glaciations. Table 6.1 summarizes the main features of the periods of evaporite deposition in these two areas.

In Section 3.3.1, we discussed the Mediterranean as an ocean in the terminal stages of its life cycle (Table 3.1, Stage 5). We must emphasize

Table 6.1

Approximate age of event (Ma ago)	Mediterranean	Red Sea
1	uplift of Gibraltar sill	
	normal marine conditions with access of cold deep Atlantic water	normal marine conditions
5	Gibraltar 'waterfall' evaporite deposition (with intermittent flooding) in shallow basins far below normal sea-level	connection with Indian Ocean and isolation from Mediterranean
	normal marine conditions	deposition of evaporites in shallow spreading basin
20	isolation from Tethyan Ocean to east	connected to Mediterranean

that evaporites are not an *essential* feature of either young or old oceans, although they are more likely to occur in them than in major ocean basins because young and old oceans tend to be narrow deep basins that can be isolated from the world ocean by either tectonic forces or falling sea-level, or both. Salt domes will occur wherever the evaporites are overlain by a sufficient thickness of other sediments (Figures 6.2 and 6.12).

6.2.6 THE MIGRATION OF CLIMATIC BELTS

You may have been wondering how warm conditions could be sustained in the Mediterranean, while ice-caps were growing at the poles. To resolve this apparent paradox, we digress briefly to discuss climatic variations and the way in which evidence from the distribution of oceanic sediment types and the biological remains they contain can be used to determine how climatic belts have expanded and contracted without changing their relative positions.

Different kinds of planktonic organisms can be very useful climatic indicators (as indeed are many land plants and animals). The remains of organisms that lived in the Quaternary glacial and interglacial periods have been used to trace the expansion of cold 'high-latitude' climatic zones towards the Equator and the concomitant contraction of temperate ('mid-latitude') zones as the ice-sheets advanced—and vice versa as they retreated. Figure 6.15 illustrates this effect, showing how present-day summer temperatures in the northern Atlantic compare with those of 18 000 years ago, at the end of the last glacial maximum.

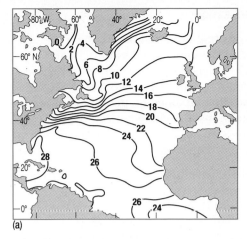

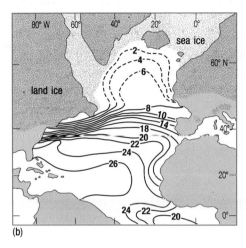

Figure 6.15 Maps comparing the distribution of surface temperatures in the northern Atlantic summer (a) at the present time and (b) 18000 years ago. Isotherms are in °C (determined in (b) from foraminiferal plankton assemblages in about 100 deep-sea cores). The margins of sea-ice in (b) are deduced from the characteristics of the sediments and from the knowledge that the sea-level was then about 100m lower than now.

QUESTION 6.12   Does Figure 6.15 suggest that the tropical climatic belt north of the Equator was significantly 'compressed', compared with its present extent, during the last glacial maximum?

On longer time-scales, of course, evolutionary changes occur in the plankton, including those which are used for temperature measurements. The lack of barriers in the oceans means that planktonic organisms appear to evolve synchronously around the globe. These evolutionary changes in the plankton, coupled with information from remains of other organisms and the sediments in which they occur, along with the magnetic reversal time-scale, provide powerful tools for interpreting the history of oceans. During the past 180Ma, events about 1Ma apart can readily be distinguished and for the last 25Ma this resolution can sometimes be as good as 100000 years or better.

In Mesozoic times, the tropical climatic belts were generally wider than at present and the poles were free of ice. As implied at the end of Section 6.2.3, this means that global sea-level must have been some 60m higher than now. The total range of sea-level fluctuations resulting from the growth and decay of ice-caps is of the order of 200m. Yet the geological record preserves evidence of numerous world-wide **transgressions** and **regressions** (advances and retreats of the sea over the land), in response to rises and falls of sea-level, over periods when the climate was consistently warm and there was little or no polar ice. We hinted at another cause of sea-level variation in Section 6.2, and we now examine it in more detail.

6.2.7   THE EFFECT OF PLATE-TECTONIC PROCESSES ON SEA-LEVEL

There are two major types of plate-tectonic process that may be important in causing rises in sea-level that lead to transgressions.

1   As you will have appreciated in answering Question 6.2(c), if the rate of production of new oceanic crust increases for any reason, there will be greater amounts of hot and therefore less dense crust standing above the general level of the surrounding deep ocean floor, and this will displace some of the water in the ocean basins onto the surrounding continents. Such an explanation has been cogently argued to account for the world-wide transgression that characterized the Upper Cretaceous (*c.* 90Ma ago) and probably implies a sea-level rise of 300–400m (see Figure 6.16). At that time, shallow epicontinental seas covered a considerably greater area than today.

2   During periods of continental break up (Figure 3.1), the overall elevation of the continents is reduced by crustal thinning (Figure 3.2) and increased loss of continental material by erosion. This erosion adds sediment to the ocean basins, and so a certain amount of water is displaced. The result is a relative rise in sea-level. Periods of prolonged world-wide continental erosion following a phase of dispersal should therefore produce marine transgressions.

Conversely, the collision of continents (as in the Alps and Himalayas) causes great thicknesses of sediments along continental margins to be uplifted and continental blocks to be thickened and isostatically elevated. The result is a fall in sea-level and regression as ocean waters withdraw into the slightly larger and deeper basins that remain.

135

Table 6.2 Major glacial periods in the last part of the geological record.

Period	Approx. time of peak
Quaternary	within the past 2 Ma
Permo-Carboniferous	c. 250 Ma
Late Ordovician	c. 450 Ma
Late Precambrian	c. 650 Ma
Upper Proterozoic	c. 900 Ma

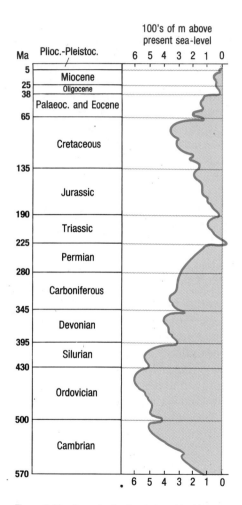

Figure 6.16 A graph of estimated world-wide sea-level during the past 570 Ma.

6.2.8 MAJOR TRANSGRESSIONS AND REGRESSIONS

Marine transgressions and regressions that occurred before the Quaternary cannot be documented in the same detail as those that occurred during the Pleistocene. There have been five well-documented ice ages over the past 900 Ma (Table 6.2), and there is no reason to suppose that the other four were not accompanied by the same kind of short-term glacial–interglacial fluctuations that characterized the Quaternary glaciation; but there is no way of demonstrating this conclusively, because the record is too fragmentary.

There is no doubt, however, that there have been larger and more long-term marine transgressions and regressions, which have taken place over millions rather than tens or hundreds of thousands of years, and evidently have nothing to do with ice ages. Classic examples of important transgressions are those of the Upper Cambrian (510 Ma ago) and the Upper Cretaceous (90 Ma ago).

One interpretation of world-wide (eustatic) sea-level over the past 570 Ma, based primarily on the sedimentary record, is shown in Figure 6.16. There are many local (isostatic) effects which may influence the picture, so you should not pay much attention to the short-term fluctuations, but concentrate on the main trends, which can be assumed to apply globally.

QUESTION 6.13 Examine Figure 6.16 and Table 6.2. Is there any obvious evidence that the major glaciations listed in Table 6.2 could have exerted a dominant control on the trends in sea-level revealed by Figure 6.16?

Continental freeboard: The concept of global **continental freeboard** is related to sea-level and its fluctuations. It may be defined as the long-term mean height of the continental surface above mean sea-level (Figure 6.17) and it is presently of the order of 0.8 km (Figure 2.4). The two main factors controlling long-term variations in freeboard are the rate of cooling of the Earth since its formation, and the rate of growth of continental crust through geological time.

The Earth is cooler now than it was in the geological past, and so the mean depth of the ocean floors is greater now than it was then, i.e. the total volume of the ocean basins has increased with time. If this increase in ocean basin volume has been balanced by an increase in the volume of continental crust, then continental freeboard will have remained constant, ignoring transgressions and regressions, whose durations are short compared with the age of the Earth. It appears that freeboard has been within ± 200 m of its present value for most of the last 2500 Ma at least. We need not concern ourselves with details of the long-term changes of freeboard, but we can look at shorter-term fluctuations.

Figure 6.17 Diagram showing the concept of continental freeboard.

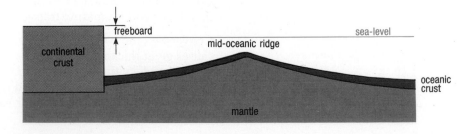

QUESTION 6.14   Present-day freeboard is estimated to be some 20m greater than 'normal' freeboard.

(a)   What does that statement mean, in relation to Figure 6.17?

(b)   How does it relate to the statement made at the end of Section 6.2.3, that sea-level would rise about 60m if all the remaining ice were to melt?

(c)   Was continental freeboard probably greater or less than 'normal' about 80–90Ma ago (Figure 6.16)?

Both sea-level and freeboard can be affected by rates of formation of oceanic crust. In the last Chapter of this Volume, we shall briefly examine other long-term aspects of the interaction between seawater and oceanic crust.

6.3   SUMMARY OF CHAPTER 6

1   Sediments of oceanic seismic layer 1 thicken from a few metres maximum at ridge crests to several kilometres at continental margins. Deep-sea pelagic sediments consist of clays and biogenic sediments formed of calcareous or siliceous skeletal remains of mostly planktonic organisms. Calcareous sediments tend to predominate on the flanks of ocean ridges, and siliceous sediments and clays accumulate in deeper parts of the ocean basins.

Metalliferous sediments resulting from hydrothermal activity form the base of the sediment sequence nearly everywhere.

2   Sediments and the organisms preserved in them provide much information about the nature and timing of events in the evolution of ocean basins. For example, the spread of a species of the planktonic foraminiferan genus *Guembelitria* in the southern oceans has helped to refine our knowledge of the history of continental separation in this region and the development of the Antarctic Circumpolar Current.

3   World-wide (eustatic) changes in sea-level are caused by changes in the total volume of water in the oceans, and by changes in the shape and volume of the ocean basins. Sea-level is presently rising world-wide, as a result of global warming after the last glaciation, enhanced by the greenhouse effect of increased CO_2 emissions into the atmosphere. Huge volumes of water stored in reservoirs and used for irrigation make the rate of rise less than it would otherwise be.

4   Local isostatic effects such as sediment loading or rebound after disappearance of ice-sheets complicate the study of sea-level changes.

5   Sea-level changes prior to the last glaciation can be reliably determined by measuring $\delta^{18}O$ values in the carbonate skeletal remains of marine organisms. The values are higher when the ice-caps are large, because water vapour (and hence the snow which builds the ice-caps) is relatively enriched in $H_2^{16}O$. $\delta^{18}O$ values in marine organisms are lower when ice-caps are small leaving the oceans comparatively enriched in $H_2^{16}O$. It is possible to calibrate $\delta^{18}O$ values in terms of sea-level changes, and water temperatures. This has enabled the growth of the Antarctic ice-sheet to be charted. Growth of the ice-sheet became especially rapid during the Miocene.

6   In the late Miocene, a combination of tectonic forces and sea-level fall due to growth of ice-sheets isolated the Mediterranean from the world ocean. The sea evaporated and thick deposits of evaporite salts were laid down. There is evidence of alternations of flooding and desiccation from about 5.5 to 4.8 Ma ago, before the barrier at Gibraltar was finally submerged, and normal conditions were re-established.

7   The growth and decay of ice-caps is accompanied by expansions and contractions of climatic belts, especially expansion of the cold polar zones. The extent of these fluctuations can be monitored by the study of organisms in deep-sea sediments.

8   Plate tectonic processes are a major long-term influence on sea-level. Periods of rapid sea-floor spreading and continental fragmentation lead to reduction in the total volume of the ocean basins, so there are widespread transgressions onto continents. Continental collision results in thickening and uplift of parts of the continental crust, causing sea-level to fall. Infill of ocean basin margins by sediments eroded from the continents contributes to sea-level rise.

9   Continental freeboard is the mean global height of the land above sea-level. It has not changed greatly over the last few billion years because continental growth has balanced the deepening of ocean basins with time. Short-term changes of sea-level cause changes of freeboard, but variations in freeboard are smaller than variations of sea-level.

Now try the following question to consolidate your understanding of this Chapter.

QUESTION 6.15   Which of the following statements are true, and which are false?

(a)   The total range of fluctuation in sea-level as a result of growth and decay of ice-caps and glaciers is of the order of 150–200 m.

(b)   Estimates of past fluctuations in continental freeboard can provide clues to the way that the geoid has varied over geological time.

(c)   The freeboard of thin continental crust will always be greater than that of thick continental crust.

(d)   Remains of *Guembelitria* in marine sediments enabled the full development of the Antarctic Circumpolar Current to be precisely dated at 20 Ma ago.

(e)   Evaporite salt sequences are always formed at stages 2 and 5 of the ocean-basin life cycle (Table 3.1).

CHAPTER 7 | THE BROADER PICTURE

In this final Chapter, we attempt to stand back and take a global perspective. The evolution of the ocean basins cannot be separated from that of the ocean waters which fill them. Changes in the shape, size and distribution of the ocean basins combine with climatic changes to influence the circulation and even the composition of seawater. These in turn control the nature and distribution of biological activity and the sediments deposited on the ocean floor. Thus, Figure 6.1 applies only to the present-day oceans, and the pattern is subtly but inexorably changing with time.

7.1 THE GLOBAL CYCLE

Chemical elements have been cycled through land, sea and air ever since the Earth acquired its fluid envelope (Figure 7.1). The mantle is the fundamental source of crustal rocks, but Figure 7.1 shows that it also contributes directly to the oceans and atmosphere by de-gassing, i.e. through the expulsion of juvenile volatile constituents (water, CO_2, HCl, SO_2 and many other gases), which are products of volcanic activity. The gaseous emanations can be recycled back through the mantle if they are trapped or chemically combined in rocks or sediments of the oceanic crust and carried down in subduction zones.

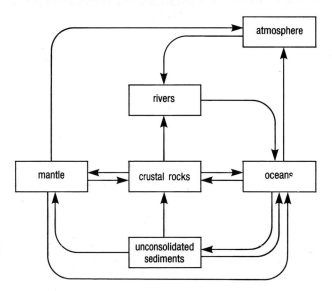

Figure 7.1 The global cycle, involving crust and mantle as well as oceans and atmosphere, illustrating the pathways eventually taken by the elements that pass through the oceans.

The cycle from continent to ocean water to ocean floor and sediments, and eventually back to continent can be summarized in four stages, as described below. At each stage, there may also be direct or indirect chemical exchange with the atmosphere.

Stage 1:   Weathering of rocks involves the removal of CO_2 (and H^+) from the atmosphere by rain and releases cations such as Ca^{2+}, Na^+, K^+ and Mg^{2+} into solution in rivers, along with HCO_3^- (bicarbonate) as the dominant anion.

The presence of dissolved CO_2 makes rainwater slightly acidic:

$$CO_2 + H_2O = H_2CO_3 \text{ (carbonic acid)} \rightleftharpoons H^+ + HCO_3^- \rightleftharpoons 2H + CO_3^{2-}$$

An important weathering reaction in sedimentary rocks is:

$$CaCO_3 + H^+ \rightarrow Ca^{2+} + HCO_3^- \quad (7.1)$$

(calcite, a common mineral in sedimentary rocks) (from rainwater) (in solution)

An example of the type of weathering reaction important in igneous and metamorphic rocks is:

$$2NaAlSi_3O_8 + 2H^+ + H_2O \rightarrow Al_2Si_2O_5(OH)_4 + 4SiO_2 + 2Na^+ \quad (7.2)$$

(albite, a common mineral in igneous and metamorphic rocks) (from rainwater) (kaolinite, a clay mineral which remains in the weathered rock) (silica, partly in solution) (in solution)

Other reactions of the general form acid + base = salt + water are common in weathering. The acid is provided mainly by carbon dioxide (CO_2) in solution. CO_2 is the fourth most abundant gas in the atmosphere (after N_2, O_2, and Ar), maintained there by volcanism, the respiratory activity of organisms and the decay and combustion of organic matter. The bases in the reactions above are supplied by the rocks, and the salts are carried away in solution, in the form of their constituent anions and cations.

What are the principal anions available in seawater to balance the cations there? What is their source?

They are Cl^- and SO_4^{2-} (Table 5.2), which are products of mantle degassing by volcanic activity (as HCl and SO_2). They are dissolved in rainwater and removed from the atmosphere, and enter the oceans either directly or via rivers. For most purposes, their concentrations are negligible in rain and river water, where bicarbonate is the dominant anion. The removal of bicarbonate from solution in seawater is explained in the next stage.

Stage 2: Dissolved constituents of seawater are removed by the deposition of sediments. This involves the biogenic precipitation of calcium carbonate and silica in the tests of organisms (i.e. the removal of Ca^{2+} and HCO_3^- and SiO_2 from solution) and the sedimentation of clays, which have been slightly modified by cation-exchange reactions on their passage from river to sea.

Stage 3: We have seen in Chapter 5 how sea-floor weathering and hydrothermal activity are highly effective mechanisms for removing cations from seawater and releasing volatiles. Other reactions that remove dissolved constituents involve the sediments.

Diagenesis may be defined as the reactions that occur during the consolidation of sediment to form rock (e.g. when siliceous and calcareous deposits become chert and limestone, respectively). In the diagenesis of clays, Mg^{2+} and K^+ are removed from solution in pore waters and incorporated into the clay minerals. The concentration of these elements is maintained in the pore waters by downward diffusion from overlying seawater.

Stages 2 and 3 involve reactions that are collectively termed **'reverse weathering'**, i.e. the removal into solid phases of the dissolved constituents of seawater. The term was coined when it was supposed that these constituents came almost entirely from rock weathering on land. It is still a useful term, even though it is now recognized that a sizeable proportion of some elements in seawater comes from hydrothermal activity.

The cycle can now be completed:

Stage 4a: Consolidated sediments and underlying rocks may be uplifted directly back into the weathering part of the cycle (e.g. ophiolite complexes, uplifted continental margins).

Stage 4b: More commonly, sediments and oceanic crust may *either* be deeply buried and metamorphosed, especially where thick continental margin sequences are caught up in continental collisions; *or* they may be subducted and partly melted to form magmas. In both cases, further chemical changes occur, followed by eventual uplift or volcanic rise back into the weathering part of the cycle (Stage 1).

Figure 7.2 summarizes the four stages described above.

QUESTION 7.1   Would you say that oceanic crust is the same after travelling a few thousand kilometres across the sea-floor 'conveyor belt' as when it leaves its source at the spreading axis? Explain your answer.

7.1.1   CHANGES IN COMPONENTS OF THE CYCLE

Chemical elements have been moving through the cycles of Figures 7.1 and 7.2 (and other related sub-cycles) for the past 3.5 billion years at least. Yet there appear to have been no major changes in the composition of *mantle* or *crustal* rocks during the whole of that time—for instance, basaltic and granitic rocks throughout the Precambrian are virtually identical with basalts and granites of Tertiary age; and the same can be said of most sedimentary rocks. On the other hand, evidence from the sedimentary record suggests that both *oceans* and *atmosphere* were somewhat different from what they are now, for much of geological time, and almost certainly before about 1000Ma ago.

While it is likely that before 1000Ma ago the relative proportions of the major cations Ca^{2+}, Mg^{2+}, K^+ and Na^+, and of Cl^- in seawater were at least qualitatively similar to what they are today, there was more HCO_3^- and less SO_4^{2-}. Sulphur was mainly combined as insoluble sulphide (S^{2-}) rather than soluble sulphate (SO_4^{2-}). Iron in the form of Fe^{2+} (which is much more soluble than Fe^{3+}) was a more important constituent of seawater (and freshwater) than it is today. It behaved in the same way as Ca^{2+} and Mg^{2+}, being precipitated as carbonate and silicate minerals.

In other words, the Earth's surface was once a more reducing environment than it is now, because the atmosphere contained more carbon dioxide and less oxygen. The decrease in the atmospheric $CO_2:O_2$ ratio is related to the evolution of the biosphere. Photosynthesizing plants

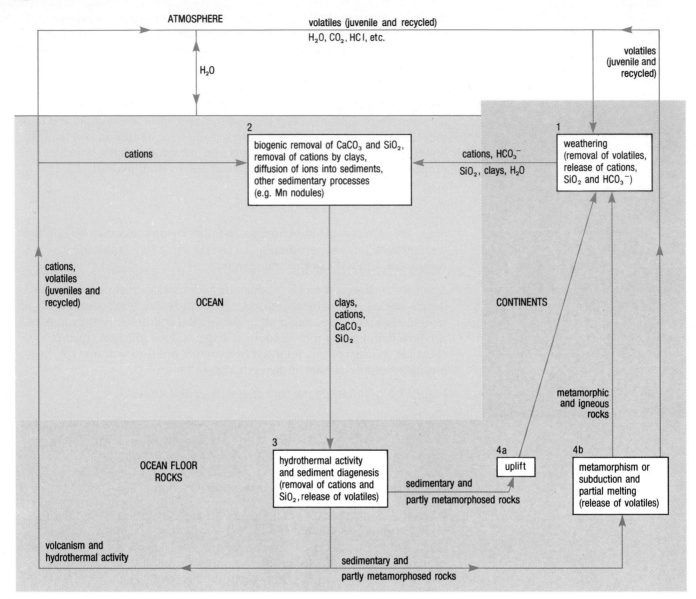

Figure 7.2 Details of the global cycle.

fix CO_2 and release oxygen, but oxygen concentrations increased only gradually over geological time, because oxygen is used up again, both in the oxidation of organic matter (respiration and decay) and in oxidation reactions during the weathering of rocks (e.g. oxidation of Fe^{2+} to Fe^{3+}). It is not known when the present level of atmospheric oxygen was reached, but it has probably been maintained near its present level for about the past 300 Ma, throughout which time massive tropical forests have existed and land-based life become increasingly diverse.

Evidence that seawater also may not have changed much over the same period comes from the preservation in marine sediments of particular organisms (such as certain echinoderms), whose basic physiology would not tolerate substantial departures from the range of composition (and temperature) that characterize the modern oceans.

This does not rule out the possibility either of a gradual change in composition with time, or some fluctuations about a long-term average composition. It would be rash to assert, for instance, that if you were to analyse a sample of water from the open oceans of, say, 150 Ma ago (in the Jurassic), you would necessarily find that the salinity and the relative proportions of major ions (let alone trace constituents) fell within the range of values typical of present-day oceans. You already know that world-wide climatic changes lead to variations in the amount of water locked up in ice-caps. When the total volume of water in the ocean changes, it must affect the salinity—although the relative proportions of the various dissolved constituents should remain the same. On the other hand, the evaporation of large volumes of seawater from restricted basins leads to the differential precipitation of dissolved constituents.

7.1.2 SOME EFFECTS OF SHORT-TERM CHANGES

We can look at those last two examples in a little more detail, and estimate their effect on the world ocean as a whole. Both can be classified as short-term events, for they have durations of the order of 10^5 years, which are almost trivial when compared with the geological time-scale.

The Messinian event (Section 6.2.5) was a crisis for the Mediterranean; but what was its effect on the world ocean?

If we assume for simplicity a total of 1 km thickness of evaporites over about four-fifths of the area of the Mediterranean floor (i.e. 2×10^6 km^2) (cf. Question 6.10), that would give us a total volume of $2 \times 10^6 \times 1 \times 10^9 = 2 \times 10^{15}$ m^3. At an average density of 2×10^3 kg m^{-3}, that is about 4×10^{18} kg of salts in the Mediterranean evaporites.

QUESTION 7.2 (a) The oceans have 1.4×10^{21} kg of water, and each kilogramme contains 35 g of dissolved salts. What is the total mass of salts in the oceans?

(b) What percentage of your answer to (a) is represented by the Mediterranean evaporites?

Clearly, on a global scale, the Messinian salinity crisis caused the removal of a significant fraction of the total salt in the world ocean. However, remember that this removal did not happen all at once: it was spread over a total of about 700 000 years, and salts continued to be carried into the open oceans by rivers.

In Section 6.2.3 you read that the Quaternary fluctuations of sea-level were the result of around 50×10^6 km^3 of water being alternately withdrawn from and returned to the oceans.

QUESTION 7.3 (a) Approximately what proportion is that of the total amount of water (1.4×10^{21} kg) in the oceans? (The average density of seawater is 1025 kg m^{-3}.)

(b) Would you expect the average salinity of seawater to increase or decrease during a glacial maximum?

7.1.3 THE STEADY-STATE OCEAN

Evidence from different parts of the global cycle (Figures 7.1 and 7.2) has led marine scientists to conclude that the oceans are in a long-term steady-state condition: the rates of input and removal of the majority of dissolved constituents are generally in balance over periods that are very long (tens to hundreds of millions of years) compared with those necessary for oceanic mixing (a few thousand years).

In other words, the rate of removal by 'reverse weathering' and subsequent processes (stages 2 to 4 in Figure 7.2) must be closely matched by the rate at which dissolved constituents are liberated in weathering reactions and returned to the oceans (stage 1 in Figure 7.2).

On average, over periods of tens of millions of years, for every mole of potassium or magnesium removed from seawater (during hydrothermal activity or diagenesis, for example) another mole must be released by weathering to find its way back to the oceans—otherwise the rates of supply and removal within the oceanic sub-cycle would not balance.

But it should be clear from Figures 7.1 and 7.2 that we cannot confine ourselves simply to the balance between weathering and 'reverse-weathering' processes. Figure 7.2 shows that juvenile constituents continue to be added to the cycle, notably H_2O, CO_2 and HCl in volcanic gases. Magmatic rock materials from the mantle, erupted at ocean ridges, affect the composition of seawater through sea-floor weathering and hydrothermal activity. It is difficult to know just how much of this contribution from the mantle is truly juvenile and how much is being recycled (over periods of up to hundreds of millions of years) from sediments and crustal rocks (plus seawater trapped within them). Such recycling would occur during metamorphism or following subduction into the mantle at destructive margins. The actual *net* addition of new constituents, however, is generally considered to be very small in comparison with the amounts being circulated in the cycles shown in Figure 7.2.

We can look at that last point in a little more detail, considering just one constituent of seawater—silica, SiO_2—by way of example. You know that hydrothermal solutions become considerably enriched in silica as a result of reactions between heated seawater and basalts (Table 5.2). Some of that silica is precipitated from cooling hydrothermal solutions, as quartz within cracks and fissures. Some of it is also taken up by silica-secreting organisms and is later deposited in siliceous sediments, and therefore becomes part of the weathering–'reverse weathering' cycle. This silica originated in basalt derived by partial melting of peridotite in the upper mantle, and so it would appear to be added to seawater from a juvenile source. Some of it (the quartz filling cracks and fissures) is likely to be carried back down into the mantle at subduction zones, but at least a fraction of the siliceous sediments becomes part of the sediment–seawater cycle (Figure 7.1), via stage 4a in Figure 7.2. If this fraction is a significant proportion of the amount released into seawater from the mantle, via basalt, then there should be a long-term accumulation of silica in these surface cycles, and a long-term depletion of silica in the mantle, and hence in basalts. However, the silica content of peridotite and basalt has not changed measurably throughout geological time, nor is there any evidence that the amount of silica has steadily increased with time in sediments and metamorphic rocks. It can be argued that because silica is so abundant it is a poor example to choose—but there is no evidence of major changes in abundance of other elements that are added to surface cycles in this way.

It is also worth considering the removal of elements from seawater, using magnesium as an example. You know now that magnesium is removed from seawater and incorporated into basalt by hydrothermal activity (Section 5.2.2). As most of the altered basalt is eventually carried down into the mantle at subduction zones, this process would appear to result in a loss of magnesium from the oceans to the mantle. By analogy with our previous argument, therefore, sediments and seawater would become steadily depleted in magnesium and the mantle would become progressively enriched in magnesium. Again, there is no evidence that this has happened, either for magnesium or for other elements that behave in the same way. Thus, the upper mantle must have been involved in the global cycling of elements through the oceans (as indicated in Figure 7.1) for the greater part of geological time.

The long-term geochemical stability of the Earth's outer skin is particularly impressive when you look at it in the context of the obvious physical and biological changes that have affected the shape and appearance of continents and ocean basins.

QUESTION 7.4   How might the steady-state ocean respond (a) to a decrease in the supply of dissolved Ca^{2+} from rivers, and (b) to an increase in the supply of dissolved Mg^{2+} from rivers?

7.2   SOME RATES COMPARED

In this final Section, it is useful to look back and make some comparisons between the rates of the various processes you have been studying.

Rates of formation of oceanic crust and of plate movement are very slow compared with the rates at which hydrothermal systems operate. At a modest spreading rate of $2\,cm\,yr^{-1}$, a piece of newly formed oceanic crust will have moved $200\,km$ in $10\,Ma$. In that time, the equivalent of the entire volume of the ocean waters will have circulated through the oceanic ridge system, through literally thousands of black and white 'smokers', the lifespan of each of which is only a few thousand years (limited by the rates of penetration of cracking fronts in hydrothermal systems, which are of the order of metres per year).

Rates of reaction in hydrothermal systems are very rapid: in laboratory experiments where seawater and basalt are heated together, the complete removal of magnesium from seawater into basalt is achieved in a matter of a week or two at 300°C, but several months at lower temperatures (and sea-bed pressures).

The high rates of reaction ensure that any significant change in the conditions under which the reactions occur should be rapidly reflected in the hydrothermal solutions emanating from spreading axes. Such changes are rare, and there is much evidence of long-term consistency in the composition and temperature of the vent solutions. This is perhaps only to be expected, for temperature and pressure do not vary much along spreading axes, nor is there a great deal of variation in the chemical composition of oceanic basaltic rocks.

The vent solutions emerge with velocities that range from tens of centimetres to metres per second. However, those rates of motion apply only to the narrow upflow zone of hydrothermal plumbing systems (Figure 5.5). Water moves through the much broader downflow zone at rates comparable with the advance of the cracking front, i.e. not more than a few centimetres per day. This is consistent with the periods of time required for the chemical reactions to reach equilibrium in hydrothermal systems.

When we compare the rates at which these crustal processes operate with rates of global sea-level fluctuation, we find that rates of sea-level change are not greatly different from those of plate movements. The present-day rise in sea-level is 1–2 mm yr^{-1}, but a glance at Figure 6.8 shows that this is atypical, at least so far as glacial–interglacial intervals are concerned.

Can you see that over much of the time since the last glacial maximum sea-level has risen at a rate of almost 1 cm yr^{-1}?

Figure 6.8 shows that sea-level rose about 80 m between about 17 000 and 7 000 years ago, which is 8×10^{-3} m yr^{-1}, or 8 mm yr^{-1}.

Localized or isostatic changes of sea-level can be much more rapid: the formation of a raised beach, for example, can involve vertical movements of several tens of cm per year—these result from tectonic forces along particular parts of continental margins.

Most of this Volume on ocean basins has been concerned with 'solid Earth' components of Figures 7.1 and 7.2, but it should already be obvious that the oceans are a very important part of the global cycle. Other Volumes in this Series are concerned principally with sub-cycles that involve the oceans and the atmosphere; although interactions with the solid Earth will, of course, be considered when appropriate.

7.3 SUMMARY OF CHAPTER 7

1   The oceans are an important part of the global cycle of chemical elements. Cations (positively charged ions) are supplied to the oceans mainly as a result of atmospheric weathering of terrestrial rocks, and partly from hydrothermal activity. Anions (negatively charged ions) to balance these in the freshwater environment come from the atmosphere (mainly CO_2), but in the oceans the cations are balanced mainly by chloride and sulphate anions, whose main origin is volcanic de-gassing of the Earth's interior.

2   'Reverse weathering' is the removal into solid phases of dissolved constituents in seawater. The resulting sediments and metamorphosed crustal rocks may eventually be returned to the continents by some combination of accretion, subduction, collision, metamorphism, volcanism and uplift.

3   The atmosphere and oceans have had approximately their present compositions for at least the past few hundred Ma. Previously, the Earth's surface was a more reducing environment, because there was less oxygen and more CO_2. Evolution of the biosphere has played a major role in reduction of the CO_2:O_2 ratio.

4   The oceans are believed to be in a long-term steady-state condition: the supply and removal of dissolved constituents are in balance. The same

appears to apply to the crustal and mantle components of the global cycle, for there has been no major change in their chemical composition through geological time.

5   Rates of movement of water through hydrothermal systems are orders of magnitude greater than the movements of lithospheric plates. Huge volumes of water are cycled through oceanic crust and huge quantities of chemical elements are exchanged in the time that it takes for a piece of crust to move a couple of hundred km from the ridge axis. Rates of global sea-level rise and fall are comparable with those of horizontal plate movements.

Now try the following questions to consolidate your understanding of this Chapter.

QUESTION 7.5   Direct measurements of the permeability of the rubbly layer 2A of oceanic crust indicate rates of 'groundwater' movement of approximately $6 \times 10^{-7} \, \text{m s}^{-1}$. How long does it take water to move 1 m through the rocks?

QUESTION 7.6   In Section 5.5 you read that over 10% of the calcium supplied to the oceans comes from hydrothermal sources. Yet the amount of water circulating through ocean ridges ($c. \, 10^{14} \, \text{kg}$) each year is less than 0.5% of the annual supply of river water to the oceans ($c. \, 300 \times 10^{14} \, \text{kg yr}^{-1}$) which provides most of the rest of the calcium. Can you explain the contrast?

APPENDIX THE STRATIGRAPHIC COLUMN

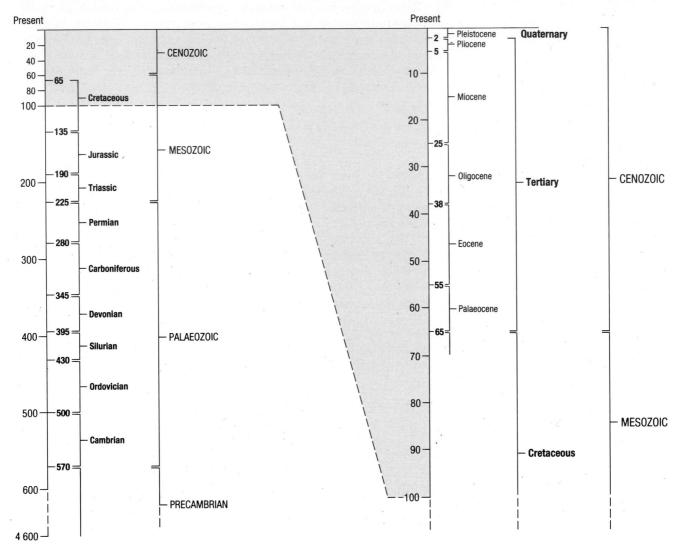

Appendix   The stratigraphic column. Ages are given in millions of years (Ma).

SUGGESTED FURTHER READING

ANDERSON, R.N. (1986) *Marine Geology—A Planet Earth Perspective*, John Wiley & Sons. The most geologically oriented book on this list. Wide-ranging and especially useful for background material, but lacking discussion of the chemical implications of hydrothermal processes.

DEVOY, R.J.N. (ed.) (1987) *Sea Surface Studies: A Global View*, Croom Helm. An interdisciplinary review of the state of the knowledge of sea-level changes and its causes, and of the oceanographic, environmental, biological and geological implications.

KENNETT, J. (1982) *Marine Geology*, Prentice-Hall. A more geological view of many of the topics in this Volume. It also introduces other aspects such as ocean currents and nearshore processes.

MENARD, H.W. (1986) *The Ocean of Truth: A Personal History of Global Tectonics*, Princeton University Press. The story of marine geological exploration during the 1950s and 1960s which led to the plate tectonics revolution, and of the scientists who participated.

RONA, P.A., BÖSTROM, K., LAUBIER, L., SMITH, K.L. JR. (eds.) (1983) *Hydrothermal Processes at Seafloor Spreading Centers*, NATO Conference Series IV, Marine Sciences, Vol. 12, Plenum Publishing Co. A comprehensive review of the topic, well illustrated with maps and photographs.

SCIENTIFIC AMERICAN (1977) *Ocean Science*. A collection of articles from the *Scientific American* magazine which report various developments in oceanography during the 1950s, 60s and 70s. Some of these are noticeably dated, but worth looking at.

ANSWERS AND COMMENTS TO QUESTIONS

CHAPTER 1

Question 1.1 The reliability with which *positions* can be determined depends mainly upon the navigational technique used. The eastern continental shelf of the United States is about 150 km wide, so the edge of the continental shelf is beyond the range of line of sight navigational methods. If either a satellite or a *Decca*-type navigation system were used (Table 1.1), the positional errors for both (a) and (b) would be in the order of 100 m, which is not important on the horizontal scale of Figure 1.8. If only celestial navigation were available, then the position of both features could be out by as much as a few km.

To obtain *depth* data, time differences have to be converted into distances, and the principal potential for error is in using the wrong sound velocity for the conversion. Moreover, low-resolution echo-sounders will not always give reliable results in areas of variable topography (Figure 1.9(b)). However, (a), the edge of the continental shelf is relatively featureless and the water is shallow, so that the profile is probably reliable to within a few metres.

When we consider (b), the 'bump' on the continental slope, we have to appreciate the great vertical exaggeration of Figure 1.8. This feature is actually about 1 km across and not much more than 50 m high, so the slopes are fairly gentle. It seems probable that here also the error is unlikely to be more than a few metres.

Question 1.2 If the satellite repeats its track every few days (preferably at intervals that are not multiples of the tidal cycle), then changes which result from the behaviour of varying winds, currents and tides can be monitored at regular intervals. Once the extent of these variations is known for particular locations or regions, the necessary corrections can be made.

Question 1.3 The horizontal scale suggests that the wavelength of geoid variations, as deduced from sea-surface height, is of the order of several tens to a few hundreds of km. The vertical scale suggests that the amplitude of these variations is between about 1 and 2 m.

CHAPTER 2

Question 2.1 (a) About 70–71% of the Earth's surface lies below sea-level, according to both the histogram and the hypsographic curve.

(b) The scale of the hypsographic curve shown in Figure 2.4(b) makes it impossible to tell precisely, but a rise in sea-level of only 100 m would flood an extra 5–10% of the Earth's surface. There would be a drastic and dramatic shift in the position of shorelines, which are ephemeral enough features on an historical time-scale and are much more so in terms of geological time.

(c)   The total vertical relief is 8.85km + 11.04km, from the hypsographic curve. This is only about 0.32% of the Earth's mean radius, which corresponds roughly to the sort of relief you find on the skin of an orange.

Question 2.2   (a)   The principle of isostasy requires that thinned continental crust 'floats' lower than thick crust. The analogy with wood blocks in water is a good one: the top of a thin wooden block will be closer to the water surface than that of a thick block and its freeboard will be less.

(b)   The transition between continental and oceanic crust is clearly not at the coastline. It must lie somewhere beneath the region between the edge of the continental shelf and the upper part of the continental rise (*cf.* Figures 2.5 and 2.6).

Question 2.3   (a)   (i) The continental shelf is consistently quite narrow, rarely more than 50km across. The shelf break can be gradual or abrupt (as with aseismic margins), but (ii) the continental slope is convex throughout its considerable vertical extent of up to 8km. The continental rise is missing completely, because of the trench at the foot of the slope.

(b)   In general, continental slopes are steeper than along aseismic margins, though the upper part of profile 1 has a gradient of only 1 in 35. The slope in profile 2 is about 1 in 8 (*c*.7°).

(*Note*: It is important to bear in mind the great vertical exaggeration in these profiles; the vertical exaggeration in Figure 2.7 is × 50 and in Figure 2.9 it is × 25.)

Question 2.4   Successively *younger* slices of sediment are scraped from the descending plate and are added to the *bottom* of the pile making up the inner trench wall. It follows that the oldest slices are at the top.

Question 2.5   (a)   The Mid-Atlantic Ridge is profile (a), and the East Pacific Rise is profile (b).

(b)   The Mid-Atlantic Ridge obviously has much rougher surface relief.

(c)   The average gradients are less than 1°, which is generally considerably less than those of continental margins, except for the continental shelf (Figure 2.7 and related text).

(d)   The Carlsberg Ridge should resemble profile (a) more than (b), because its spreading rate is similar to that of the Mid-Atlantic Ridge.

Question 2.6   Figure 2.18 shows a clear shelf break, a continental slope with a gradient of about 1 in 100, and a well-defined continental rise, beginning about 200km from the shelf break at about 2km depth and grading down into the abyssal plain some 600km out and 4km down. The plain itself grades into abyssal hill topography at about 1500km from the shelf edge and at about 5.5km depth. The aseismic margin of a major ocean is indicated. Therefore the most obvious answer is the Atlantic— and, in fact, this profile is of the Cape Verde abyssal plain off West Africa.

Question 2.7   (a)   3400 km in 43 Ma will give an average rate of movement of:

$$\frac{3400 \times 10^3}{43 \times 10^6} = 0.079\,\text{m yr}^{-1} = 7.9\,\text{cm yr}^{-1}$$

(b)   At the same rate of movement, the age at the northern end of the chain would be approximately:

$$\frac{1900 \times 10^3}{7.9 \times 10^{-2}} + 43\,\text{Ma} = 24 + 43\,\text{Ma} = 67\,\text{Ma}$$

(c)   The kink could be explained by a change in the direction of movement of the Pacific plate at about 43 Ma ago. However, more data from these and other chains would be needed for confirmation.

(d)   The duration of the period concerned is 72–43 Ma = 29 Ma. To account for 1900 km of movement during this period, the average rate would be:

$$\frac{1900 \times 10^3}{29 \times 10^6} = 0.066\,\text{m yr}^{-1} = 6.6\,\text{cm yr}^{-1}$$

Question 2.8   To form a ridge, the hot-spot volcanism would have to be virtually continuous rather than episodic, as in seamount/island chains.

Question 2.9   (a)   There are three or four features suggesting segments of a probably aseismic ridge parallel with the Ninety-east Ridge. Neither this ridge nor the basins between it and the Ninety-east Ridge fit well with the bathymetric chart.

(b)   The Diamantina Trough is associated with a linear east–west geoid low immediately to the north, which extends far to the east, along the southern margin of Australia, well beyond the eastern end of the trough on the chart.

(c)   The correspondence is good, and the features all have geoid anomalies in excess of +2 m.

(d)   The more important fracture zones show up as linear boundaries between geoid anomaly features. This is particularly clear on the East Pacific Rise (Figure 2.23(c) and (d)).

(e)   The Tonga–Kermadec Trench appears as a very deep low in the geoid. In fact, though the contours do not make this clear, its maximum depth in the geoid is 20 m. There appear to be ridges in the geoid, about 3 m high, parallel to and on either side of the trench. These are in fact artefacts, introduced by the filtering technique used to remove long wavelength features in the geoid. They do not represent real bathymetric ridges. Similar artefacts show up near the Java Trench and probably the Diamantina Trench in Figure 2.23(a). Artefacts are associated with all the real bathymetric features in these geoid anomaly maps, but as their magnitude is only about 15% of that of the associated real feature, they are usually too small to affect the contours at this scale.

Question 2.10   Slow-spreading ridges are characterized by strong topographic relief compared with fast-spreading ridges (Figure 2.11), reflecting the effect of crustal cooling and subsidence with age (Figure

2.13). For a given distance from a ridge, crust formed at a slow-spreading ridge will be older and therefore deeper than crust formed at a fast-spreading ridge. The geoid in the vicinity of a spreading ridge is correlated with the topography of the ridge: the more pronounced the ridge, the larger the anomaly. In general, therefore, the amplitude of a geoid anomaly over a ridge is inversely correlated with spreading rate.

Question 2.11 The depth of the sea-floor must be 4000m. This corresponds to an age of about 15–20Ma, using Figure 2.13. The assumptions that must be made are that the volcano formed at the same time as the portion of oceanic crust upon which it rests, and that local isostatic adjustments can be ignored so that it can be assumed to have subsided at the same rate as the crust.

If the volcano formed at a ridge crest, its summit would have been about 1.5km below sea-level. This is far below the limits of wave action, so we should not expect a flat top due to wave action. The same argument applies even more strongly if the volcano is younger than the age estimated above.

Question 2.12 (a) False. The age–depth relationship means that since sea-floor subsides at the same rate for a given age, whatever the spreading rate, then crust generated at a slow-spreading ridge has to travel *less far* before it has subsided to a given depth. Figure 2.11 (and related text) shows that the gradients on fast-spreading ridges are gentler than those on slow-spreading ridges.

(b) Could be either true or false: the evidence is not sufficiently conclusive. We have given only two examples in the text; one supports the model, the other is less clear. It is an attractive idea, but there are insufficient data to be certain about the majority of linear island chains.

(c) True. As the crust gets older, more and more sediments accumulate on it.

(d) True. The principle of isostasy (analogy with wood blocks) requires thinned continental crust to be less elevated than continental crust of normal thickness. The only circumstance in which a large area of continental crust of normal thickness could be below sea-level would be if it were forcibly held down in some way.

(e) Partly true. Some sediment is subducted and much is also scraped off onto the inner wall.

Question 2.13 (a) The sea-floor on the north of the fracture zone is deeper and therefore older than that on the south side. Since it is older, it must have spread farther from the ridge at which it was formed. This means that the ridge is offset to the left. Note that the sense of offset is the same whether you imagine yourself looking northwards or southwards along the ridge.

(b) The age–depth relationship shows that the age of sea-floor with a depth of 5000m is about 47Ma. We know that this depth is reached at a distance of 940km from the ridge. As we know both the age and the amount of spreading, we can calculate the average spreading rate:

$$\text{rate} = \frac{\text{amount of spreading}}{\text{age}} = \frac{940 \times 10^3}{47 \times 10^6} = 2.0 \times 10^{-2} \, \text{m yr}^{-1} = 2.0 \, \text{cm yr}^{-1}$$

(c)   Spreading rates of around $2\,\mathrm{cm\,yr^{-1}}$ and less are classified as 'slow'. In terms of its spreading rate, this ridge resembles the Mid-Atlantic, Carlsberg, Central Indian and South-west Indian Ridges.

(d)   To work this out, we need to know how far the neighbouring sea-floor, north of the fracture zone, is from the segment of the ridge at which it formed. We are told that the depth here is 5500m. According to the age–depth relationship, this corresponds to an age of about 69 Ma. As spreading rate is almost certain to be identical on both sides of the fracture zone, we can calculate the amount of spreading:

$$\text{amount of spreading} = \text{rate} \times \text{age}$$
$$= 2.0 \times 10^{-2} \times 69 \times 10^{6}$$
$$= 1.38 \times 10^{6}\,\mathrm{m} = 1380\,\mathrm{km}$$

So, north of the fracture zone we are at a distance of 1380km from the ridge, whereas south of the fracture zone we are at a distance of 940km from the ridge. The offset on the ridge is simply the difference between the two:

$$1380 - 940 = 440\,\mathrm{km}$$

A completed map of the area is shown in Figure A1.

(e)   There is a considerable amount of scatter in the data from which the age-depth curve is derived, and the curve itself is beginning to flatten off as the lithosphere cools, making it difficult to relate depth precisely to age. Based on the age–depth relationship only, there could be errors of $\pm$ 5 Ma, or about $\pm$ 10%, in the age estimate. The estimate of spreading rate (b) could therefore be in error by at least $\pm$ 10%. When we come to *compare* the ages of the sea-floor on either side of the fracture zone, the imprecisions in the age–depth curve become more critical. If the age estimates for the two sites were each out by 5 Ma in opposite senses, then the age difference would be in error by 10 Ma, and the estimate of the ridge offset would be out by 200km. Magnetic anomaly data would help a lot here.

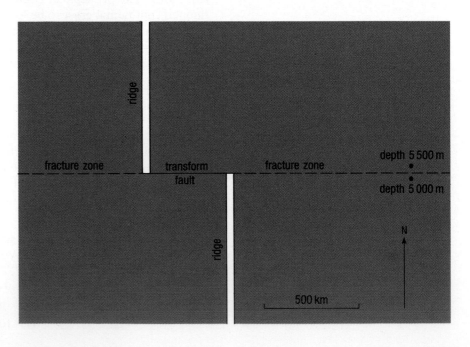

Figure A1   Completed version of Figure 2.25 (for use with answer to Question 2.13).

CHAPTER 3

Question 3.1 (a) Tethys has been shrinking since the Jurassic (170Ma) to the present because of the rotation of Eurasia and Africa towards each other, and the northward movement of India. The Mediterranean Sea and probably the Black and Caspian Seas are all that is left of Tethys.

(b) It should be apparent from Figure 3.1 that different parts of the Atlantic Ocean began to open at different times. From this Figure, we can see that the southern Atlantic (between S. America and Africa) and central Atlantic (between N. America and W. Africa) began to open between 170Ma and 100Ma ago, but the northern Atlantic (between Greenland and Europe) did not begin to spread until sometime between 100Ma and 50Ma ago.

(c) Between 170 and 100Ma ago, according to Figure 3.1.

(d) 170Ma to 100Ma, because Gondwanaland had already begun to break up by 100Ma ago: the Atlantic was open in the south and the Indian Ocean had begun to open as Africa separated from Australia and Antarctica. During the period 100Ma to 50Ma ago, India travelled very rapidly northwards towards Asia, Australia began to move away from Antarctica, and Panthalassa, of which the Pacific is the remnant, shrank.

(e) The Atlantic and Indian Oceans continued to expand, and India collided with Asia, causing the formation of the Himalayas. The Pacific continued to shrink, and island arcs developed along its western margins.

Question 3.2 The pole of relative rotation between two plates can be located by reference to spreading ridge segments and transform faults separating the plates (Figure 2.14). In the Red Sea, the axial rift segments are offset from one another but are roughly parallel to the length of the Red Sea (Figure 3.3(a), Figure 3.4). The pole must lie somewhere close to the extension of a line running along the length of the Red Sea. The Red Sea tends to be widest in the south-east and we think that the opening of the Red Sea is propagating from south-east to north-west, so we can assume that the spreading rate is greatest in the south-east. As spreading rate decreases towards the pole, the pole must lie to the north-west. There are no clearly defined transform faults within the Red Sea, but the fault along the Gulf of Aqaba/Dead Sea line is effectively a transform margin (a conservative plate boundary) and so must lie approximately along an arc of a small circle about the pole. In fact, the pole is thought to lie somewhere between Sicily and Crete.

Question 3.3 (a) The patterns on Figure 3.6 are not easy to unravel, but there is clear evidence of a *southward* increase in age (52 to 65 to 80Ma) east of the ridge; and of a *northward* increase (65 to 80Ma) west of the ridge.

(b) They were developed well before the present South-east Indian Ridge. Symmetrical spreading about this ridge began about 35Ma ago, and in a NNE/NE to SSW/SW direction, rather than north–south.

(c) The ridge which generated the crust east of the Ninety-east Ridge must have lain to the north, because ages decrease in that direction. There is no crust younger than 40–50Ma there, so the ridge must have been subducted down the Java Trench.

Question 3.4 The oldest crust in the Mediterranean is in the east, and it is likely that the thickest sediments are there, too. The western Mediterranean has much younger crust. The older the crust, the deeper it lies (Figure 2.13), so the western Mediterranean should have more 'buoyant' crust that the eastern. If the sea-surface (the geoid) mimics the configuration of the ocean floor (Figure 1.17), then we would expect it to be more depressed in the east, as it is in fact.

Question 3.5 From Figure 3.6, the widths of ocean floor age bands south of India are greater for the period before about 45Ma ago than they are for subsequent periods. Spreading rates and therefore rates of movement of continental blocks would have been greater before 45Ma ago than after.

Question 3.6 By simple proportion, using a ruler, the ratio of width of sea-floor from 0–52Ma for the East Pacific Rise to that for the Mid-Atlantic Ridge is close to 5:1.5 or 3.33. If the average spreading rate for the Mid-Atlantic Ridge was $2\,\mathrm{cm\,yr^{-1}}$ over this period, then for the East Pacific Rise it was 2×3.33 or nearly $7\,\mathrm{cm\,yr^{-1}}$.

(*Note:* These are spreading rates per ridge *flank*; the total spreading rate will be double these values.)

CHAPTER 4

Question 4.1 The speed of sound in water is $c.\,1.5\,\mathrm{km\,s^{-1}}$ (Section 1.1.2), which is less than that in rock (Figure 4.1). Water will occupy fractures in rocks on the sea-bed, and so will reduce the speed of sound (the seismic velocity) through fractured rocks relative to unfractured rocks, in which there is no water to 'retard' the sound waves.

Question 4.2 In the case of the Mid-Atlantic Ridge, if the half-spreading rate (the rate at which a plate moves away from the axis) is $2\,\mathrm{cm\,yr^{-1}}$, then the total rate of formation of new oceanic crust is twice this, i.e. $4\,\mathrm{cm\,yr^{-1}}$, or $0.04\,\mathrm{m\,yr^{-1}}$. At this rate, 1m of spreading is accomplished, on average, in a period of $1/0.04 = 25$ years. So one new dyke can be expected about every 25 years. The East Pacific Rise is spreading four times faster, so if the dykes are the same width then they will be generated four times as often, about once in every six years. As you will see later (Section 4.1.2), the episodic nature of spreading means that these figures can be regarded only as long-term averages, on time-scales of 100000 years and longer.

Question 4.3 (a) The largest faults will be at the edges of the rift valley where the crust is uplifted to form the east and west walls.

(b) The most likely locations are the hills near the centre of the rift valley, such as Mount Venus. However, as you will discover, volcanic rocks cover the whole rift-valley floor.

Question 4.4 Figure 4.11(a) shows two pillow tongues which probably formed in the manner described in Figure 4.4. The flat area surrounding the pillows may be sediment, but there is no evidence of sediment resting upon the pillow tongues.

Figure 4.11(b) shows a surface formed of numerous pillow lavas with a thin veneer of sediment.

Figure 4.11(c) shows a fissure that cuts the rift-valley floor. This fissure is cutting pillow lavas and sediment.

Figure 4.10(d) shows rock debris (probably broken-up pillow lavas). This typically forms at the base of steep fault scarps.

Question 4.5 (a) Figure 4.5 shows four transform faults offsetting the Mid-Atlantic Ridge axis. There are no other offsets. The lengths of axis between these transform faults are 11km, 36km and 47km, the average of which is 31km. Figure 4.18 shows the East Pacific Rise axis interrupted by four transform offsets and eight overlapping spreading centres. There are thus 12 segments in about 1000km, making an average segmentation interval of about 83km.

(b) Your results suggest that segmentation occurs at more widely spaced intervals along fast-spreading ridges (e.g. the East Pacific Rise) than along slow-spreading ridges (e.g. the Mid-Atlantic Ridge). However, remember that you have very little data to go on (only three segments on the Mid-Atlantic Ridge). It would be very unwise, therefore, to read too much into this. In fact, based on much more data, some marine geologists have suggested that the segmentation interval *does* increase with spreading rate, but there is a lot of scatter within the data and this correlation has not been widely accepted.

Question 4.6 (a) All it means (if true) is that in the long term, overlapping spreading centres stay put, but may migrate or oscillate over short distances about some mean position every few hundred thousand years.

(b) Rock samples dredged from the western limb of the overlapping spreading centre will have been derived from the magma chamber beneath the topographic high immediately to the south; samples from the eastern limb would have come from the magma chamber beneath the topographic high to the north. We should expect the samples to show some subtle differences in their chemical and mineralogical characteristics.

Question 4.7 Hydrostatic pressure tends to prevent the escape of dissolved gas from magma erupted under water, and any gas that does escape will be compressed into smaller vesicles than at atmospheric pressure. The sudden release of these high pressures can sometimes have dramatic effects. For example, blocks of vesicular lava dredged from deep water off southern California spontaneously burst apart when exposed to atmospheric temperatures and pressures on board ship—the area has become known as 'Popcorn Ridge' in consequence. However, vesicles are not commonly found in submarine lavas beneath about 500m in depth.

The predominance of pillow lavas over volcanic ash in the lower parts of oceanic volcanoes can be explained by high hydrostatic pressures at the sea-floor which prevent gas escaping explosively. Chances of a 'steam explosion' are much reduced because the high pressures not only raise the boiling point of water but also inhibit the expansion of any steam that might form.

CHAPTER 5

Question 5.1   (a)   The *observed* measurements of heat flow across spreading axes differ increasingly from the *predicted* values as the ridge crest is approached. The theoretical curves assume that heat is transferred only by conduction. However, as observations show this to be less than predicted there must also be transfer of heat by convection of seawater within the crust.

(b)   The shaded areas on Figure 5.6 represent the 'heat loss deficit', the *difference* between predicted conductive rate of heat loss (broken line) and the observed conductive heat flow. It is this difference which is transferred out of the oceanic crust by convecting seawater.

(c)   Both curves have an exponential form, because increasing depth (Figure 2.13) and decreasing conductive heat flow (Figure 5.6) are closely related in models of cooling lithospheric plates.

(d)   The fact that the four components of Figure 5.6 are so similar shows that this pattern must be typical of all the world's oceans. On all four diagrams, observed heat flow matches prediction from about 70 Ma onwards (*cf.* Figure 2.13). Figure 3.6 suggests that oceanic crust up to about 70 Ma old occupies about one-third of the area of the world's oceans.

Question 5.2   (a) (i)   The variation in the analyses of Table 5.1 enables us to be sure only that two constituents are added to basalt from seawater during hydrothermal metamorphism, namely MgO and H_2O.
(ii) Similarly, we can be sure only of one constituent that is consistently leached from rocks, namely CaO. There is some indication that SiO_2 and K_2O are leached to varying degrees. The data in Table 5.1 suggest that Na_2O, TiO_2, iron oxides and MnO may be either added to the rocks or removed from them (*cf.* Section 5.2.2).

(b)   The answer lies in fault scarps. Faulting in axial regions forms escarpments roughly parallel with ridge trends. Transform faults form scarps and clefts across the trend of ridges. These features can be many hundreds of metres high and their lower parts will expose deeper layers of oceanic crust at the surface (Section 2.4 and Figure 2.16).

Question 5.3   (i)   Table 5.2(b) shows that hydrothermal solutions have a pH of around 4, and so must be more acidic than ordinary seawater which has a pH of around 8.   (ii)   They are also more reducing, because they contain sulphide (as H_2S) rather than sulphate in solution, as well as increased iron and manganese, which are more soluble in reducing than in oxidizing conditions.

Question 5.4   If we assume a cracking front penetration rate of $2 m yr^{-1}$, it would take about 2500 years for the water to penetrate to a depth of 5 km, i.e. this is the time required for the rock at 5 km depth to have become cool and brittle enough to crack. Thus the lifespan of both magma chambers and vent systems should be of the order of millennia (10^3 years), unless the magma chamber is replenished by fresh melt from below within that time.

Question 5.5 You may have been tempted, by analogy with the conductive anomaly at the axis, to say that low heat flow represents convective upflow. However, off axis the situation is different. Downflowing water carries heat away, cooling the rock and reducing its heat flow. Upflowing water brings with it the heat which it gained elsewhere and warms the rock up, increasing its heat flow. Note that there is still an overall deficit compared with the theoretical conductive heat flow, because some heat is transported out of the crust (into the seawater) by convection.

Question 5.6 The seismic compressional wave velocities of unmetamorphosed and metamorphosed oceanic crustal rocks are not sufficiently different to enable definite discrimination to be made between them using seismic information alone. For example, gabbros are indistinguishable from amphibolites and young basalts are indistinguishable from greenschists. Figure 5.9 shows only data for seismic compressional wave (P-wave) velocities, but similar overlaps occur between shear wave (S-wave) velocities.

Question 5.7 (a) Substituting in equation 5.1:

$$F = \frac{2 \times 10^{20}}{4 \times 10^3 (300-2)} \approx 1.7 \times 10^{14} \, kg \, yr^{-1}$$

(b) The total mass of ocean water divided by the rate calculated in (a) gives:

$$\frac{1.4 \times 10^{21}}{1.7 \times 10^{14}} \approx 8 \times 10^6 \text{ years}$$

Question 5.8 Table 5.3 shows that hydration (the amount of H_2O incorporated in the rock) in sea-floor weathering is less extreme than in hydrothermal alteration; that sea-floor weathering causes potassium to be added to the rocks; and that oxidation causes Fe_2O_3 to increase considerably relative to FeO. The other effects (alteration of glass and feldspars, deposition of manganese oxide) are less obvious in Table 5.3.

Question 5.9 Figure 5.7 shows that the isotherms are very close together for temperatures less than about 300°C. The crustal thickness in which the temperature range is within the stability field of chlorite is much greater than that appropriate to zeolite stability. That is why chlorite is so much more abundant.

Question 5.10 $6 \, km^3$ per year is $6 \times 10^9 \, m^3 \, yr^{-1}$, which is $6 \times 10^9 \times 2.8 \times 10^3 \, kg \, yr^{-1}$ of basalt, i.e. about $1.7 \times 10^{13} \, kg \, yr^{-1}$. The average water:rock ratio is then given by:

$$\frac{1.7 \times 10^{14}}{1.7 \times 10^{13}} \approx 10:1$$

This average can be expected to conceal considerable variations, because more rock per unit time is generated at faster spreading axes than at slower spreading ones, so water:rock ratios may well be greater at slow- than at fast-spreading ridges. Magmatism is segmented along axes, and the intensity of hydrothermal circulation must vary accordingly.

Question 5.11 (a) 1300 p.p.m. magnesium in 10^{14} kg of water is 1.3×10^{11} kg of magnesium, the amount lost from seawater to rocks annually.

(b) The figure in our simple calculation is the same as for the river flux. Hydrothermal systems must provide *the* major sink for magnesium in the oceans: the concentration of magnesium in seawater is kept constant because a quantity equal to all the magnesium supplied annually by rivers is removed in hydrothermal systems each year.

Question 5.12 (a) False. Sea-floor weathering will occur wherever seawater at bottom temperatures is in contact with oceanic crustal rocks.

(b) Arrant nonsense.

(c) True. The first sediments to accumulate on igneous crust are deposited at ridge crests and are enriched in Fe and Mn (and some other metals) discharged at hydrothermal vents (Figures 5.4 and 5.5). Any sediments deposited thereafter will overlie these metal-rich sediments.

(d) Could be either true or false. Crust is thinner at transform intersections, but magma chambers peter out towards them (Chapter 4). The temperature *difference* across a given *thickness* of crust may be similar. The thermal gradient depends also on the nature of hydrothermal circulation.

(e) False. Looking at Figure 5.7 it is likely that only the uppermost layer of rubbly lava, layer 2A, is saturated with cold seawater. Below that, water is beginning to heat up in downflow zones.

(f) True. The only source of ^{3}He of any consequence is mantle outgassing from ocean-ridge axes. Although the ^{3}He escapes to the atmosphere, its highest concentrations will be in the oceans, though it is diluted all the time by ^{4}He, which is being produced wherever rocks contain U or Th.

CHAPTER 6

Question 6.1 (a) Sediments deposited at site 283 were silts and clays throughout the Palaeocene and the first part of the Eocene. These are likely to have been relatively near-shore sediments, deposited from rivers and redistributed by waves, tides and currents. Only later in the Eocene did the sea become wide and deep enough at this point for siliceous organic remains to dominate the sediments, undiluted by continent-derived sediments.

(b) Site 281 is the only one of the four sites where sediments were deposited on continental rather than oceanic crust.

(c) Calcareous (carbonate) sediments are accumulating at all four sites today, and this is consistent with the information in Figure 6.1. Deep-water sedimentation at sites 281 and 282 has been carbonate-dominated since the Miocene.

Question 6.2 (a) The most important single factor controlling the volume of water in the oceans is the balance between the freezing and melting of ice, especially in the polar ice-caps, but also in mountain glaciers.

(b)   A 1°C increase in average temperature would cause the volume of the oceans to increase by a factor of 2.1×10^{-4} or 0.021%. The average depth of the oceans is 3.7km (Figure 2.4). Neglecting any resulting change in surface area (which will be very small in comparison with the total surface area), an average temperature rise of 1°C throughout the oceans would cause the depth to increase by a factor of 2.1×10^{-4}, due to thermal expansion of the water. This is a rise of:

$$3.7 \times 10^3 \times 2.1 \times 10^{-4} = 7.8 \times 10^{-1} \approx 0.8\,\text{m}$$

So, ignoring ice-sheets, global temperature changes will cause sea-level changes of the order of 1m. This does not seem much in relation to the depth of the oceans, but when we are considering changes of the order of millimetres (see the text that follows), then temperature changes could become an important factor.

(c)   The depth to which oceanic crust will sink on cooling after it spreads away from a constructive plate margin depends upon its age (Figure 2.13) and not upon the distance it has moved since its formation. It follows that oceanic crust produced at a fast-spreading ridge will have travelled further from the axis before it sinks to its equilibrium level. A fast-spreading ridge will therefore generally be broader and higher than a slow-spreading ridge. So, other things being equal, the ocean basin will be shallower and more water will be displaced onto continental shelves worldwide (*cf.* answer to Question 2.10).

Question 6.3   Figure 6.6(a) shows that the overall (secular) rise in sea-level for the 70-year period 1890 to 1960 was from about 20mm to 150mm on the vertical scale, a total rise of 130mm, giving an average rise of 1.8–1.9mm per year.

Question 6.4   (a)   Line (c) on Figure 6.7 shows a rise of about 65mm in 50 years giving an average rate of 1.3mm per year. Line (d) shows a rise of about 100mm in 50 years, or 2.0mm per year. This second figure is close to the rate you worked out in Question 6.3. Remember that Figure 6.6 shows sea-level at *one* site, whereas Figure 6.7 is a global average.

(b)   Even to maintain the rate of rise in sea-level at the reduced value of about 1.3mm per year, there must be a progressive increase in storage of water and/or use in irrigation schemes. That is because the global warming will ensure a continuing rise of sea-level. The process may ultimately be self-defeating, because at least some of the increased amount of water vapour in the atmosphere, produced by evaporation from all these endeavours, will precipitate directly back into the oceans.

Question 6.5   Sea-level is still rising within a belt extending from southern Britain through the North Sea and into the Netherlands and northern Germany. However, it is falling around most of Scandinavia, because this region is still rebounding isostatically from the release of the ice-loading after the most recent glaciation. This rebounding has outpaced the general rise in sea-level caused by the ice melting.

Question 6.6   (a)   Equation 6.1 tells you that a positive $\delta^{18}O$ value means the $^{18}O{:}^{16}O$ ratio in the sample must be greater than that in the standard. It therefore indicates enrichment in ^{18}O in the sample relative to the standard (SMOW in this case).

(b)   A negative $\delta^{18}O$ value must by the same argument indicate relative depletion of ^{18}O.

(c)   As organisms growing in cold water tend to incorporate more of the heavy ^{18}O isotope into their skeletons, the $^{18}O{:}^{16}O$ ratio must be higher in organisms growing in cold water. We should therefore expect to find higher $\delta^{18}O$ values in the skeletons of such organisms.

(d)   Evaporation enriches the ^{16}O isotope in water vapour, relative to the water left behind. There is therefore less ^{18}O in ice than in seawater and $\delta^{18}O$ values should be negative in polar ice.

Question 6.7   (a)   The temperature of surface waters varies much more than that of water in the deep oceans, where the variation in temperature throughout the Quaternary is unlikely to have exceeded a degree or two, as at the present day. Because the oxygen-isotope ratio in calcareous skeletal remains depends partly on the proportion of ^{16}O that is incorporated into polar ice-caps and partly on the temperature of the water in which the organisms grow, this second variable is largely eliminated by measuring isotope ratios of remains of organisms that grew where there was very little temperature variation. Benthic foraminiferans fulfil this criterion; but so, to some extent, do planktonic foraminiferans in high latitudes.

(b)   From Figure 6.10, and with the information that a 0.1 per mil change in the $\delta^{18}O$ value corresponds to a 10m change in sea-level, it is possible to deduce (i) a rise of about 100m for the last post-glacial interval, and (ii) a slightly higher rise (about 120m) for the preceding interval.

(c)(i)   The last major fluctuation in sea-level was not especially noteworthy because the last glaciation was not significantly different from its predecessors.

(ii)   The rises in sea-level that follow glaciations are evidently more abrupt than the falls, which seem to be characterized by oscillations superimposed on an overall downward trend.

Question 6.8   (a)   The first definite signs of ice-transported debris in the sediments appeared in the late Oligocene, and the surface-water temperatures were then of the order of 7°C.

(b)   There is a fairly abrupt steepening of the downward trend in the mid-Miocene, and this could reasonably be correlated with an acceleration of the growth of the Antarctic ice-sheet.

Question 6.9   If the net rate of water loss of $3.25 \times 10^3 km^3 yr^{-1}$ were maintained, since the initial volume of water in the Mediterranean is $3.75 \times 10^6 km^3$, the time taken to dry out would be:

$$\frac{3.75 \times 10^6 km^3}{3.25 \times 10^3 km^3 yr^{-1}} \approx 1.15 \times 10^3 \text{ years}$$

Question 6.10   The thickness of evaporites with the specified volume and surface area would be:

$$\frac{6.5 \times 10^4 km^3}{2 \times 10^6 km^2} = 3.25 \times 10^{-2} km, \text{ or } 32.5m$$

Question 6.11 (a) This part of the question is designed to ensure that you recognize the general correlation: the higher the $\delta^{18}O$ value, the more ice there was in glaciers and ice-caps, and hence the lower the sea-level. Average global temperatures were probably lower also (*cf.* Section 6.2.3) which would tend to reduce the rate of evaporation, but nevertheless evaporite deposition occurred during the periods of lower sea-level, when the Mediterranean was isolated and so could dry up due to evaporation.

(b) Each of the numerous peaks and troughs in the raw data indicates fluctuating ice volumes and hence fluctuating sea-levels. Short-term rises and falls of global sea-level would provide the necessary inundations to replenish the supply of seawater in the Mediterranean and produce a new 'crop' of evaporites.

(c) The time-scale varies somewhat: there are about 10 peaks in the interval 5.89–5.35 Ma, and about another 10 in the interval 4.77–3.88 Ma. The total time interval in the first case is about half a million years, in the second about 900000 years. The time-scale of fluctuations thus varies from about 50000 to 100000 years. This approximates the time-scale of fluctuations in Figure 6.10. However, it is crucially important when making such comparisons to bear in mind the resolution of the sampling. In Figure 6.14, the sampling interval was every 70 cm of sediment, corresponding to about 20000–25000 years; in Figure 6.10, sampling was about every 10 cm, corresponding to about 5000–6000 years. Thus, the record in Figure 6.10 resolves details which are lost in Figure 6.14.

(d) The Messinian salinity crisis lasted from about 5.5 Ma ago (when the overall lowering of sea-level and tectonic uplift combined to isolate the Mediterranean) to about 4.8 Ma ago (when the barrier at Gibraltar was finally submerged), a total of 0.65 Ma. However, there was an interval between about 5.3 and 5.2 Ma ago when normal marine conditions intervened between the two periods of evaporite deposition.

(e) Falls in sea-level were no longer able to isolate the Mediterranean once the barrier at Gibraltar was opened finally, because the connection had become too deep either through tectonic effects or by erosion from inflowing water.

(f) In Section 6.2.4 you read about the growth of the Antarctic ice-sheet, which is the major repository of ice in the world. It is likely that most of the sea-level fluctuations in the Miocene and later times were due to changing ice volumes in Antarctica.

Question 6.12 Remarkably enough, no. The distribution of surface isotherms (lines of equal temperature) is clearly not the same in detail in both diagrams, but the zone of warm water in equatorial and tropical latitudes (up to about 30°N in Figure 6.15) was evidently not much narrower at the height of the glaciation than it is now. Further north, however, the isotherms were formerly much more crowded together as a result of the southward expansion of the high-latitude cold belt. It is the mid-latitude climate belt that was compressed, not the tropical belt.

Question 6.13 There is no obvious evidence of any such control—indeed rather the reverse: there were regressions after the Ordovician and Permo-Carboniferous glaciations, when polar ice-caps shrank. The very extensive transgression in the Cretaceous cannot be associated with the decline of any glaciation—and the ensuing regression (c. 70 Ma ago) could not have been caused by the growth of polar ice-caps, either. These major transgressions must have resulted from plate-tectonic processes.

Question 6.14 (a) The statement means that the mean elevation of the land above mean sea-level is 20 m greater now than it is normally.

(b) If all the ice melted, the volume of water in the oceans would increase, and sea-level would rise. Freeboard would decrease but *not* by as much as 60 m. That is because freeboard is an *average* elevation of the land above the sea.

(c) Figure 6.15 suggests that 80–90 Ma ago, sea-level was some 300–400 m higher than at present. Continental freeboard must therefore have been appreciably less than it is now, and probably well below 'normal'. The widespread shallow epicontinental seas of that time were mentioned in Section 6.2.7.

Question 6.15 (a) True. The rise in sea-level since the last glacial maximum is at least 90 m (Figure 6.8), and an extra 60 m could be expected if the rest of the ice melted (Section 6.2.3).

(b) False. The statement is nonsense. Freeboard fluctuations can by their very nature only be estimated from indirect evidence. Variations in the geoid over geological time could only be very crudely estimated from past distributions of continents and oceans.

(c) False. The requirements of isostasy are such that (by analogy with wood blocks in water), thin continental crust must 'float' lower than thick crust (*cf.* also Questions 2.2(a), 2.12(d)) and so would have a smaller freeboard.

(d) False. *Guembelitria* has helped to establish when open water conditions developed all round Antarctica with a probable wind-driven surface current, but it is a planktonic organism and cannot be satisfactorily used to determine when the current became fully developed at depth.

(e) False. We emphasized at the end of Section 6.2.5 that such sequences *can* be formed at these stages, but this does not invariably happen.

CHAPTER 7

Question 7.1 Your answer must be no, especially after reading Chapter 5. Considerable quantities of elements have been added to the crust *en route*, e.g. Mg and K. Large amounts of other elements have been removed, notably Ca. These and other elements have also been redistributed, so that, for example, it is likely that large amounts of iron and manganese (and perhaps copper and other ore metals) have been removed from within the crust and redeposited near or on the surface, sometimes in concentrations of economic importance. Above all, huge volumes of water have become fixed in the crust, in hydrous minerals such as clay minerals, zeolites and chlorite. These changes occur most rapidly nearest the spreading axes.

Question 7.2 (a) 1.4×10^{21} kg water with 35 g salts per kg means that the total quantity of salts is

$$1.4 \times 10^{21} \times 35 \times 10^{-3} \text{kg, i.e. } c.5 \times 10^{19} \text{kg}$$

(b) The Mediterranean evaporites represent

$$\frac{4 \times 10^{18}}{5 \times 10^{19}} \times 100,$$

or 8% of the salt in the world oceans.

Question 7.3 (a) 50×10^6 km^3 is 50×10^{15} m^3.

This is $50 \times 10^{15} \times 1025$ kg $= 5.125 \times 10^{19}$ kg $\approx 5.1 \times 10^{19}$ kg.

The total mass of water in the oceans is 1.4×10^{21} kg, so the glacial fluctuation is:

$$\frac{5.1 \times 10^{19}}{1.4 \times 10^{21}} \times 100 \approx 3.6\%$$

(b) As there is less *water* in the oceans during glacial maxima, we might expect the average salinity to be greater.

Question 7.4 (a) Less calcium entering a steady-state ocean implies less calcium removed from it or more added by another means. There are various ways in which this is likely to be achieved: decreased precipitation of calcium carbonate to form skeletal shells of marine organisms; increased dissolution of biogenically precipitated $CaCO_3$; and increased supply of calcium from hydrothermal activity.

(b) More magnesium entering a steady-state ocean implies more magnesium removed from it. Increased magnesium concentration in sediment pore waters would result in increased removal by diagenetic reactions (Figure 7.2); more magnesium would also be removed from seawater during hydrothermal circulation.

Question 7.5 The answer is given by the expression:

$$\frac{1 \text{m}}{6 \times 10^{-7} \text{ms}^{-1}} \text{ or } 1.67 \times 10^6 \text{s}$$

That works out at just over 19 days, corresponding to a rate of about 5 cm per day, which is consistent with rates to be expected in the downflow zone of hydrothermal systems, as outlined in Section 7.2.

Question 7.6 The reason is that the concentration of calcium in hydrothermal solutions is much greater than that in river water (although you have not been provided with data, you know that freshwater is not saline). A small amount of concentrated solution makes a proportionately greater contribution than a large amount of a dilute solution.

ACKNOWLEDGEMENTS

The Course Team wishes to thank the following: Dr. Martin Angel and Dr. Andy Fleet, the external assessors; Mr. Mike Hosken for advice and comment on the whole Volume; Captain Iain F. Kerr and Dr. Charles Turner for their help in the compilation of Chapters 1 and 6 respectively; Dr. Graham Jenkins and Dr. Sandra Smith for information on specific points.

The structure and content of the Series as a whole owes much to the experience of producing and presenting the first Open University course in Oceanography (S334) from 1976 to 1987. We are grateful to those people who prepared and maintained that Course, to the tutors and students who provided valuable feedback and advice and to Unesco for supporting its use overseas.

Grateful acknowledgement is also made to the following for material used in this Volume:

Figure 1.1 G. F. Bass (1974) *A History of Seafaring based on Underwater Archaeology*, Thames and Hudson by permission of the Metropolitan Museum of Art; *Figure 1.2* N. J. W. Thrower; *Figure 1.3* Naval Museum, Madrid, and Arxiu Mas, Barcelona; *Figure 1.4(a)* E. Linklater (1972) *The Voyage of the Challenger*, John Murray Ltd. by permission of the Rainbird Publishing Group; *Figures 1.4(b), 1.16 and 1.17* Institute of Oceanographic Sciences/NERC; *Figure 1.8* K. K. Turekian (1976) *Oceans*, Prentice-Hall Inc.; *Figure 1.11* NASA; *Figure 1.13* N. G. Dukov, Institute of Oceanography, Varna, Bulgaria; *Figure 1.14* National Geophysical Data Center; *Figure 1.18* courtesy of the National Remote Sensing Centre, Farnborough; *Figure 1.19* National Science Foundation; *Figures 1.21, 4.5, 4.6, 4.9, 4.11 and 4.12* Woods Hole Oceanographic Institution; *Figure 2.1* X. Le Pichon *et al.* (1973) *Developments in Geotechnics 6: Plate Tectonics*, Elsevier; *Figures 2.8 and 2.10* C. A. Burk and C. L. Drake (1974) *The Geology of Continental Margins*, Springer-Verlag; *Figure 2.12* H. D. Needham and J. Francheteau (1974) in *Earth and Planetary Science Letters*, **22**, No. 1, North Holland Publishing Co.; *Figure 2.13* J. G. Sclater and D. P. McKenzie (1973) *Geological Society of America Bulletin*, **84**, No. 10, Geological Society of America; *Figure 2.18* A. S. Laughton (1959) in *New Scientist*, **1**, New Science Publications; *Figure 2.19* B. G. Heezen and C. D. Hollister (1971) *The Face of the Deep*, Oxford University Press; *Figure 2.21* G. B. Dalrymple *et al.* (1973) in *American Scientist*, **61**; *Figure 2.23* T. H. Dixon and M. E. Parke (1983) in *Nature*, **304**, Macmillan Journals; *Figure 3.3(a)* E. Bonatti (1985) in *Nature*, **316**, Macmillan Journals; *Figure 3.4* E. T. Degas and D. A. Ross (eds) (1969) *Hot Brines and Recent Heavy Metal Deposits in the Red Sea*, Springer-Verlag; *Figure 3.6* J. G. Sclater and B. Parsons (1981) in *Journal of Geophysical Research*, **91**, B12; *Figure 3.7* R. G. Gordon and J. M. Gurdy (1986) in *Journal of Geophysical Research*, **91**, B12; *Figure 4.4(a)-(c)* J. G. Moore (1975) in *American Scientist*, **64**; *Figure 4.4(d)* J. R. Heirtzler and W. B. Bryan (1975) in *Scientific American*, **233**, copyright © 1975 W. H. Freeman and Co.; *Figure 4.8* Institute of Oceanographic Sciences; *Figures 4.13–4.16* R. D. Ballard and T. H. van Andel (1977) in *Bulletin of the Geological Society of America*, **88**; *Figures*

INDEX

Note: page numbers in italic refer to illustration captions

abyssal
    hill(s) 39, 40, *40*, 151
        province 39
    plains 28, 31, 39, *39*, 40, *40*, 50, 53, 114,
        151
accretion 146
acids 99, 101, 112, 140
actinolite 98, 112
active margins 29, 49
    see also Pacific-type margins; seismic
        margins
Afar region 89, *89*
Africa *29*, 37, 46, 53, 58–9, 64, 66, 73, 129,
    133, 151, 155
African plate 73
age
    -depth relationship 35–6, *35*, 37, 41, 50,
        95, 121, 153–4
    of ocean floor 26, 36, 41–2, *60*, 61, 63–5,
        86–7, 95, 116, 118, 153–4, 156, 161
    of oceans 60–6
    of seamounts 50, *89*
Agulhas Plateau 46
albite 98, 112, 140
Alpine
    Fault (New Zealand) 38
    – Himalayan mountain chain 26
Alps 65, 135
altimetry 14, 18, *18*, 19, 24, 43–4, 50
Alvin 23, *74*, 76, 77
American plate 73
amphibolite 97–8, 108, 159
ancient Egyptian river craft *5*
Andes 32, *33*
anhydrite (calcium sulphate) 100, 102, 109,
    112
anions (–) 139–40, 146
Antarctic(a) 117–18, 127–9, *128*, 137, 155,
    162–4
    Circumpolar Current 117–20, *118*, 129,
        137–8
    plate 32
Arabian plate 58–9, 129
Arabian Sea 58
aragonite (CaCO₃) 114
Arctic 86
aseismic margins 29–31, *30*, 49, 151
    see also Atlantic-type margins; passive
        margins
aseismic ridges 32, *37*, 43, 50, 152
ash 88, 115, 157
Asia 155
    Minor 65
Asmara volcano *89*
asthenosphere *25*, 26, *27*, 28, 31, 70, *71*, 84,
    85, 90
Atacama desert 32
Atlantic
    Ocean 5–6, 26, 28, 30, 37–8, *39*, 43, 50,
        52–3, 61, 65–6, *67*, 79, 88, 114, 130–4,
        135, 151, 155
    –type margin 29, 49
        see also aseismic margins; passive
            margins
atmosphere 121
atmospheric pressure 20, 24, 121, *121*, 157

'Austral Gulf' 117, 119–20
Australia 64, 117–18, 155
axial
    crest 83–4, *83–4*, 90, 92–3, *93*, 105–6,
        106, 111, 114, 116, 137, 153, 158, 160
    deeps 58, 108
    rift valley 11, 57–8, 105, 109, 155
Azores 73

back-arc basin 33, 61–2, *63*, 64, 86, 90, 107,
    111
Balkans 65
barite *see* barium sulphate
barium 109, 112
    sulphate (barite) 92, 102, 112
basalt 26, 58, 81, 85, 87, 90, 92–3, 96–9,
    101, 106, 111–13, 119, 141, 144–5,
    158–9
    weathered 113
basaltic
    dyke *69*
    magma 34, 38, 55, 70
    melt 70, 78
    pillow *69*, 78
basins 27–8, 70, 111, 114, 121, 131–8, 161
    birth of 55–9, *56*, 116
    evolution of 52–5, *54*, 117, 120, 139
    major 61–5
bathymetry 120
    echo-sounding 11, 12, *12*, 15, 24, 76, *76*,
        78, 83, *83*, 89
    line sounding 11, *11*, 14, *15*
    maps/charts 11, 13–14, *14*, 23, *32*, *33*, *34*,
        36, 43, *47*, 73–4, *78*, 83, 152
    radar 20–1, *21*
    satellite 18–21, *18*, *20*, 43–9, 50, 65
benthic foraminiferans 125, *125*, *132*, 133,
    162
benthic organisms *117*, 125, 127
bicarbonate (HCO₃⁻) 139
biogenic sediments 56–7, 114, 137, 140,
    165
Black Sea 130, 155
'black smokers' 92, *92*, 100–5, *103*, 108,
    110–13, 116, 145
block faulting 35
boron (B) 100
bottom water temperatures 92, *92*, 97,
    101–2, 105, 107, 111–13, 125
brine 107–8
Britain 161
bromine (Br) 100
Brunhes epoch *125*
bulk chemical analyses 98

calcareous sediments 114–15, 126, 137, 140,
    160, 162
calcite (CaCO₃) 102, 109, 114–15, 140
calcium (Ca) 96–7, 100, 109, 112, 139, 141,
    145, 147, 158, 164, 165
    carbonate (CaCO₃) 109, 114–15, *117*, 124,
        126, 140, 165
    sulphate *see* anhydrite
Cambrian Period 136
canyons, submarine 15, 17, 27, 30

carbon (C) 100
    dioxide (CO₂) 121–2, 137, 139–42, 146
    monoxide (CO) 111,113
carbonates 41, 132, 137, 140–1, 160
carbonic acid (H₂CO₃) 140
Caribbean *60*, 61, 65
Carlsberg Ridge 11, 34, *35*, 63, 151, 154
Caspian Sea 155
cations (+) 139–41, 146
Central Indian Ridge 48, 154
Challenger 8, 9, 11, 14, 21
chemical reactions 92, 96–101, 106–7,
    139–44, 147
chert 140
chloride (Cl⁻) 101, 140, 146
chlorine (Cl) 100
chlorite 98, 107, 112–13, 159, 164
Cibicdoides kullenbergi 132
clay
    minerals 68, 112, 114–15, 140–1, 164
    pelagic 115, 137
climate 32, *54*, 111, 124, 139, 143
climatic belts 63, 115, 118, 134–5, 138,
    163
coastlines 114, 120–1, 150–1
Columbus, Christopher 5, 7
compressional wave velocities *108*
conduction 94–6, *95*, *106*, 111, 158–9
conservative plate margins 38, 50, 155
    see also transform faults
constructive plate margins 25, 26, 28, 34,
    49, 161
    see also spreading axes
continental
    blocks 120, 135, 156
    crust 21, *25*, 26, 29–30, 49, 50, 55–6,
        58–9, 63, 65, 86, 120, 136, 138, 151, 153,
        160, 164
    freeboard 136–8, *136*, 137, 151, 164
    margins 29–33, 34, 37, 39, 49, 53, *56*, 61,
        63, 64, 114, 116, 128, 135, 137, 141,
        146, 151
    rise 29, 31, *31*, 37, *39*, 40, *45*, 52–3, *56*,
        114, 116, 151
    shelf 13, 15, 28, 29–30, *31*, 49, 52–3, *56*,
        114, 116, *116*, 124, 129, 150–1
    slope 13, 15, 29–31, *31*, *39*, 52–3, *56*, 114,
        116, 150–1
continents 28, 30, 33, 37, 49, 52–3, 61, 64,
    117, 135, 138
convection 94–5, *94*, 104–7, *106*, 109, 112,
    158–9
Cook, James 5
copper (Cu) 23, 164
    sulphide 102
Cosa, Juan de la: map 7, *7*
cosmic rays 110
cracking front 104, 146, 155
Crete *19*
Cretaceous
    Peedee Formation 126
    Period *54*, 64–6, 87, 136–6, 163–4
Crozet Islands 45
crust 21–4, 26, 50, 55, *57*, 58, 63, 64, *72*,
    85–6, *85*, 90, 100, 105–6, *139*, 147
    see also continental crust; oceanic crust

crustal
  abnormalities 86–7
  rifting 53, *56*, 65
  stretching 29–30, 55, *56*, 59, 65
    see also crustal thinning
  subsidence 29–30, 152–3
  thinning 29–30, 49, 55, 58–9, 90, 135,
    151, 160
    see also crustal stretching
currents
  ocean *see* ocean currents
  tidal 24, 43, 150, 160
Cyprus 92

Decca 10, 150
deeps 58, *58*, 65, 108
Deep Sea Drilling Project (DSDP) 22, 23,
  24, 87, 108, *119*, *128*, 178
delta ^{18}O (δ^{18}O) 126–7, 132–3, *132*, 137,
  161–2
depth-measurement: echo-sounding 11–13,
  12, *13*, 24
density of water 110, 143
destructive plate margins 25, 26, 31, 32, 49,
  53, 61, 144
  see also ocean trenches
diagenesis 140–1, 144, 165
Diamantina Trough *45*, 46, 152
dispersion of effluent 110–11
Drake Passage 117–18
dredging of samples 72, 76, 81, 87, 91, 96,
  98, 105, 108, 111, 157
drilling 22, 23, 64, 68, 81, 87, 105, 107–8,
  119, 130
DSDP *see* Deep Sea Drilling Project
dyke *69*, 70, 80–1, 90, 97–8, 108, 156
  injection 72, 81, 84, *84*

earthquakes *25*, 29, 37, 64
East African rift valley 52–3
East Pacific Rise 34, *34*, 36, 39, 48, 61–2, 66,
  70, 81, 83–6, *83*, *88*, *89*, *92*, 100–2, 105,
  110, *110*, 151–2, 156–7
echo-sounders 11, *13*, 75–6, *75*, 82–3, 150
echo-sounding 12, *12*, 14, 15, 23, 24, 48, *83*,
  89–90
  see also narrow-beam echo-sounding
Ecuador 108
Esjberg *121*, 121–2
Eltanin Fracture Zone 49, 82
Eocene *54*, 119, 128, 160
epidote 98
erosion 27, 30, 31, 41, 57, 88, 115, 119, *119*,
  132–3, 135, 138, 163
Ethiopia 89
Eurasia 53, *54*, 129, 155
Europe 38, 53, 64, 66, *123*, 124, 129, 155
eustatic sea-level changes 120, 125–5,
  136–7
evaporation 55, 120, 125–6, 130, 132, 138,
  143, 161–3
evaporites (salt deposits) 55–6, 58–9, 64–5,
  108, 116, *116*, 129–34, *132*, 138, 143,
  162–3, 165
exchange of elements (between seawater
  and rocks) 30, 92, 96, 99–101, 107,
  113, 125–6, 130–1, 140–5, 160
exploration of the oceans 5–24, 90
extensional faulting 32, *33*
  see also rifting

FAMOUS 23, *72*, 73–84, *77–8*, *83*, 84
faulting 35, 64, 72, 75, *78*, 79–81, *80*, 84,
  87, 90, 96, 104, 108, 112, 156, 158
feldspars 68, 97–8, 112, 159
fluorine (F) 100
foraminiferan 117, *117*, 124–5, *125*, 127,
  128, 133, *135*, 137, 162
fracture(s) 50, 58, 68, 84, 87, 90, 94, 96, 98,
  112, 156
  zones 36–9, *36–7*, 46, 48–50, *51*, 68, *73*,
    74, 82, 86–7, 90, 153–4

gabbro 68, *69*, 70, 81, 82, *82*, 87, 90, 97–8,
  104, 108, 159
Galapagos Rise (Ridge) 101–2, 105, *106*
geoid 18–19, *18*, *19*, 24, 64, 120, 138, 150,
  152–3, 156, 164
  anomalies 43–9, *44*, *46–7*, 152–3
geological time-scale 52, 58, 84, 91, 96, 105,
  106, 114, 120–2, *128*, 133–6, 141–5,
  147, 150, 163–4
Gibraltar 38
  sill 132, 134, 138, 163
  Straits of 130
  'waterfall' 132, *132*, 134
glaciations 120–2, 124–9, *125*, 132–5, *132*,
  136, 137–8, 146, 160–4
glass 97, 112, 159
global warming 121, 133–6, 161, 163
Globigerinoides sacculifera 125
Glomar Challenger 22, 23
GLORIA 15, *16*, 17, 39, 75, *75*
glycerine 85
Gondwanaland 53, *54*, 155
Gorda Ridge *35*
gravity anomalies 19, 24, 32, 85, *85*
'greenhouse effect' 121–2, 137
Greenland ice-cap 126, 155
greenschist 97–9, 107–8, 159
Guembelitria 117–18, *117–18*, 120, 137–8,
  164
Gulf
  of Aden 58–9
  of Aqaba/Dead Sea line 58–9, 155
  of California 30, 53, 87, 110, 157
  of Mexico *116*
  Stream 9, 114
  of Suez 58
guyots 41, *41*, 87–90

Halley, Edmund: map 5, *5*
Hawaiian
  – Emperor seamount chain 42, *42*, 43,
    49, 61
  Islands 40
heat
  flow 67, 94–6, *95*, 102, 105–6, *106*, 109,
    111, 158–9
helium 110, *110*, 112–13, 160
Henry the Navigator, Prince 5
Himalayas 26, 53, 63, 65, 135, 155
hornblende 98
hot
  brines 107–8
  spot 42, *42*, 43, 45, 50, 61, *62*, 152
  spring *see* hydrothermal vents
hydrated aluminosilicate 97
hydration 112, 159
hydrocarbons 114, 116
hydrochloric acid (HCl) 139–40, 144

hydrogen 11, 113, 139–40
  sulphide (H_2S) 116
hydrological cycle 122
hydrostatic pressure 88, 94, 96, 107,
  112–13, 157
hydrothermal
  activity 84, 94, 98, 101, 105, 137, 140–1,
    144–6, 165
  circulation 85, 90, 92–113, 116, 146–7,
    159–60, 165
  effluents 92, 110–13
  metamorphism 97, 101, 104, 108, 111–13,
    158–9
  solutions 98–102, 104, 107–8, 110–12,
    116, 144–5, 158, 165
  vents 23, 92–3, *92–3*, 96, 99–105, *103*,
    107–8, 110–12, 145, 158–60
hydrous oxide residues ('rust') 101, 164
hypsographic curve *27*, 120, 150–1

ice age 122–5, 127, 129, 136, *136*
Iceland 86, *86*, *88*, 93, *93*
imaging radar 20, 24
India 155, 156
Indian Ocean 5, 14, 18, 26, 28–9, 34, *35*,
  43–6, *44–5*, 53, 56, 58, *60*, 61, 63–5,
  107, 129, 134, 155
International Indian Ocean Expedition 14
iron (Fe) 23, 93, *93*, 97, 100–1, 112–13, 141,
  158–60, 164
  oxide 115, 142, 158
  sulphide (FeS, FeS$_2$) 100, 102
island arcs 26, 29, 33, 40, 49, 53, 61, *63*,
  64–5, 155
isostatic sea-level changes 26–7, 35, 124,
  136–7, 146, 161
isostasy 26, 41, 151, 153, 164
isotherms *103*, 159, 163
isotopic ratios 110–11, *110*, 113, 124–7, *128*,
  162

Java Trench 29, *45*, 63, 152, 155
JOIDES Resolution 23
Jurassic Period *54*, 64, 87, 143, 155
juvenile constituents of seawater 144

kaolinite 140
Kerguelen
  Islands 45
  Plateau 45

lava 38, 68, 70–1, *71*, *72*, 78, *79*, 81–2, *83*,
  85, 87–91, 157, 160
leaky transform 38, *38*, 50, 87
LIBEC 76, *76*
limestone 55, 140
Line Islands–Tuamoto seamount chain 43
line soundings 11, *11*, 14
lithium (Li) 109
lithosphere 26, *27*, 28, 31, 33–7, 41, 42, 49,
  52, 55, 61, 65, 84, 85, 95, 154
  structure and formation 67–91
lithospheric plates 26, 50, 58, 70, 81, *85*, 88,
  90, 147, 158
Louisiana *116*
Louisville Ridge 45, 48
low velocity layer 68, 156

Madagascar 29
  Plateau 46

Madeira abyssal plain *40*
Magellan, Ferdinand 5, 8
magma 38, 50, 70–1, *71*, 81, 85, 88, *88–9*,
    98, *103*, 141, 144, 159
    chamber 70, *71*, 80–2, *82*, 84–5, *84–5*,
      90–1, 94, *103*, 104–5, 157–60
magnesium (Mg) 96, 100–1, 107, 112–13,
    139, 141, 144–5, 158, 160, 164–5
magnetic
    anomalies *35*, 57, 58, *58*, *60*, 61, *62*, 65,
      73, 86, 105, 114
    field reversals 26, *27*, *59*
    stripes *35*, 57, *60*, 61, 65
    time-scale (magnetic polarity time-scale;
      magnetic reversal time-scale) *27*, *59*,
      125, 135
manganese (Mn) 93, *93*, 97, 101, 105, 109,
    112–13, 158–60, 164
    nodules 17
mantle 25–6, *27*, 32, 38, 42, 50, 55, 65, 68,
    70, 84, *85*, 90, 110, 139, *139*, 144–5, 147
    de-gassing 110–11, 139–40, 146, 157, 160
    plume 42, *42*
marginal basin 33
    see also back-arc basin
Marianas
    Island Arc 111
    Trench 87
marine transgressions and
    regressions 135–7, 163–4
Matuyama epoch *125*
Mauna Loa 40
Maury, Matthew: bathymetric map 13–14,
    14, *15*
median valley 34–5, *35*, 49, 87, 90
Mediterranean Sea 19, *19*, 38, 53, 64–6,
    129–34, *131–2*, 138, 143, 155–6, 162–3,
    165
melting of ice-caps 121–2, 124, 127, 137,
    160–1, 164
Menard Fracture Zone 48, 86
Mesozoic 135
Messinian salinity crisis 129–34, 143, 163
metal-rich sediments 17, 111, 113, 116, 137,
    160
metamorphic rocks 113, 144, 159
metamorphism 68, 72, 97–8, 104, 112–13,
    141, 144, 146
meteorites 115
methane (CH_4) 111, 113
microcontinents 29, 39, 49, 63, 65
Mid-Atlantic Ridge 11, 14, 28, 34, *34*, *35*,
    36, 38, 57, 66, 70, 72–82, *73–5*, 84, 86,
    86, 93, 104–5, 151, 154, 156–7
Middle America Trench *33*
mid-oceanic ridge *see* spreading axes
Miocene 56–7, *59*, 64, 129–33, 137–8, 160,
    162–3
Moho 68, 90
Mozambique Plateau 46
multi-deeps region 58, *58*

narrow-beam echo-sounding 76, 82–3, *83*,
    89–90
Nauru Basin 87
navigation
    acoustic 75
    celestial 10, 150
    coastal 9
    determination of longitude 6, 24
    radar 9–10

radio 9–10, 24
    satellite 10, *11*, 24, 150
    sonar 23
    visual *9*, 10
navigational accuracy 9, 15, 24, 75, 82
Nazca Plate 31
Newfoundland 92
New Zealand 46, 48
New Zealand Alpine Fault
Nile, River 130
Ninety-east Ridge 14, 43, 46, 63, 152, 155
nitrogen (N) 140
nitrous oxide (N_2O) 111, 113
normal faults 35
    see also block faulting
North America 38, 61–3, 65, 150, 155
North Atlantic Drift *see* Gulf Stream
North Sea 121–3, 161
nutrients 110, 115

ocean currents 9, 11, 20, 24, 27, 31, 41, 43,
    104, 111, 113, 114–15, 117–18, 120, 164
    see also Antarctic Circumpolar Current:
      Gulf Stream
Ocean Drilling Project (ODP) 23, 87, 108
oceanic
    crust 21–4, *25*, 26, *35*, 40–1, 47–50, 52,
      55–6, 58–9, *60*, 61, 63–8, *63*, *67*, *69*, 70,
      71, 79–81, 84, 86–8, 90, 92–6, *94*, 98,
      102, 104, 107–9, 111–14, 116–17, 120–2,
      124, 135, 137, 139, 141, 145, 147, 151,
      153, 156, 158–61
    trench *25*, 26, 29, 31, 32, 49, 50, *63*, 64
ODP *see* Ocean Drilling Project
Oligocene 117, *118*, 120, 129, 162
olivine 68, 97–8, 112
Omega 10
oozes *69*, 115
opal (silica) 100, 102, 104, 114–15, 140–1,
    144, 158
ophiolites 64–5
    complexes 64–5, *67*, 92–3, 104–5, 108,
      116, 141
Ordovician Period *136*, 163
organic matter 111, 113, 115–17, *117*, 119,
    123, 124–6, 162
overlapping spreading centre (OSC) 39, 57,
    83–4, *84*, 86, 90–1, 157
oxidation 92, 97, 99, 101, 111–12, 142,
    158–9
oxygen (O) 110, 124–5, 140–2, 146
    isotopes 124–6, *125*, 127–9, *128*, 132–3,
      132, 162

Pacific Ocean 8–9, 14, 18, 26, 27, 28, 31,
    35, 40, 41–2, *42*, 43, 45, *46–7*, 50, 52,
    53, *60*, 61–5, 86, 88–9, 101, 105, 108,
    110, 133, 152, 155
    –Antarctic Ridge 48
    –type margins 29, 49
      see also active margins; seismic margins
palaeobathymetry 36
palaeoceanography 14–38
Palaeocene 119, 160
palaeoclimate *see* climate
palaeogeography *54*
palaeomagnetism *54*, 62
Pangaea *54*
Panthalassa 53, *55*, 155
Paramour Pink 6

partial melting 33, 84–5, *85*, 112
passive margins 29, 49
    see also aseismic margins; Atlantic-type
      margins
PDB standard 126
pelagic
    clay 115, 137
    sediments 40, 50, 114–15, 133, 137
peridotite 26, 68, *69*, 70, *82*, 87, 90, 98,
    112, 144
permeability 94–6, 102, 104, 111–12, 147
Permo-Carboniferous Period *136*, 163
Peru-Chile Trench 31, 32, *32*, *33*
pH 100 158
photosynthesis 141–2
pillow lavas 68, *69*, 70–2, *72*, 76, *77*, 78, 81,
    84, 88, 90, 96–7, 108, 156–7
plagioclase feldspar 68, 97–8, 112
plankton 135, *135*
planktonic organisms 56, 114, 117, *117*,
    118, 124–5, *125*, 127, *128*, 134–5, 137,
    162, 164
Planulina wuellerstorfi 132
plate
    margins *25*, 26, 29, 32–3, 38, 61, *62*
    tectonics 21, 25, *25*, 29, 32, 52, 66, 73, 86
      and sea-level changes 133–5, 138, 146,
      163–4
Pleistocene *128*, 129, 136
Pliocene 132–3
Pluto, Mount 74, 78
polar ice-caps 114, 121–2, 125–8, 132, 134,
    137–8, 143, 162–4
pole of rotation 59, 155
Ponta Delgada 73
'Popcorn Ridge' 157
potassium (K) 96, 100, 109–10, 112, 139,
    141, 144, 158–9, 164
Precambrian *136*, 141
precipitation 97, 102, 104, 107, 112, 125,
    130, 140, 143, 161, 165
Proterozoic *136*
pyroxene 68, 97, 112

quartz 98, 100, 107, 112, *128*, 144
Quaternary Period
    ice age 122, 124–5, 127, 129, 134, 136,
      136, 162
    sea-level changes 122, 124, 127, 143

radar
    altimetry 19, *20*, 43
    –imaging 20–1, *21*, 24
    navigation 9–10
radio navigation 9–10, 24
rainwater, composition 126, 140
red clays 114–15
Red Sea 53, 55, 56–9, *57*, 65, 89, 108, 110,
    129–30, 133–4, 155
reduction 99, 111–12, 138, 141
regression, marine 135–7, 164
'reverse weathering' 141, 144, 146
Reykjanes Ridge *86*
Rhône, River 130
ridge crest *see* axial crest
ridges *see* aseismic ridges; speading axes
rift valley 35, 49, 52–3, *56*, 57–8, 65, 74–6,
    76–80, 79, 81, 90, 105, 156–7
rifting 29, 37, 52, 53, 65, 72, 79–80, 87, 118
Rio Grande Rise 43

river water, composition 109, 140, 147, 160, 165
Rockall Bank 29
rubidium (Rb) 109, 112

salinity 108, 110, 129–33, 143, 165
  see also Messinian salinity crisis
salt 55, 108, 116, 130–1, 140, 143, 165
  domes 116, *116*, 130, *131*, 134
San Andreas Fault 38
sandwaves (megaripples) *17*, 20, *21*, 30
satellites
  for bathymetry 14, 18–21, *18*, 24, 43–9, 65
  for navigation 10, *11*, 24, 150
  for remote sensing 14, 20–1, *21*
Scandinavia 124, 161
sea-floor
  age 26, 35–6, *35*, 36, 41–2, *60*, 61, 63–5, 86, 95, 116, 118, 153–4, 156, 161
  mapping 13–17, 23, *23*, 24, 26, 44, 50
  metamorphism 68, 72, 97–8, 104, 112–13, 141, 159
  rocks 64–5, 67, 76, 87, 97–8, 112, 159
  spreading 25–6, 28, 33–4, 36–7, *37*, *38*, 48–9, 52, 57, 59, 64, 66, 70, 73, 86–7, 117–18, 138
  weathering *93*, 97, 101, 104, 108, 112–13, 140, 144, 159–60
sea-level 11, 24, 88–9
  changes 19, 20, *20*, 24, 28, *89*, 114–38, 146, 161–4
Sea MARC 17, 39, *89*
seamount(s) 14, 32, 39, 40–1 *41*, 43, *45*, *47*, *47*, 50, 87, 88–90, *89*, 105
  chains 41–3, *42*, 49, 152
Seasat
  imaging radar 20, *21*
  radar altimeter *18*, *19*, 20, *20*, 43–5, *44*, *46*
seawater composition 30, 92, 96, 99–101, 107, 113, 125–6, 130–1, 140–5, 160
sediment(s) 15, 23, 27–33, 39, 49–50, 52, 55, *57*, 58–9, 64–5, 68, *69*, 87, 90, 93–6, *93–4*, 101, 106–7, *106*, 110–11, 122, 127, 129, 131, 133–5, 138, 139–40, 144–5, 157, 160
  and palaeoceanography 117–20
  supply and distribution 37, 40, 53, 76, 114–16, *115*
segmentation (of spreading axes) 36, 39, 48, 55, 82–7, 90, 157
seismic
  layers 67–8, *67*, 70, 81–2, 86–7, 90, 96–8, 101–2, 104, 108, 112–14, 137
  margins 29, 31–2, 49–50
  see also active margins; Pacific-type margins
  reflection 32, *33*, 40, 67, 81, *82*, 84, 104, *131*
  refraction 67
  velocity 67, *67*, 87, 90, 156, 159
  waves *67*, 159
serpentine 98
serpentinite 87, 98, 112
Seychelles Plateau 29
sheeted dykes 70

shelf edge 30, *31*
side-scan sonar 15, 17–18, *17*, 20, 23–4, 39, 75, 83, 89–90, *89*
  see also GLORIA; *Sea MARC*
silica (SiO$_2$) 100, 102, 107, 114–15, 140–1, 144, 158
siliceous sediments 114–15, 120, 137, 140, 144, 160
silicon (Si) 96, 100, 109, 112
sills 87
sodium (Na) 98–100, 112, 139–41, 158
Somali Current 9
sonar 15, 17–18, *17*, *22*, 23–4, 39, 75, *75*, 83, 89–90, *89*
  see also echo-sounder; side-scan sonar
sonograph 17, *17*, 75
South America *25*, 28, *29*, 32, *32*, 37, 86, 117, 129, 155
South-east Indian Ridge 48, 63, 155
southern ocean 117–18, *118*, 137
South Pole 126
South-west Indian Ridge 48, 63, 86, 154
Spanish Armada 20
spot sounding 14
spreading
  axes 26–8, 36, 38–9, 41, 48–51, *51*, 53, 61, 63–5, 68, *76*, 80–7, *82–3*, 90, *94*, 96, 104–5, 109–10, 112–13, 114, 120–1, 141, 145, 152–3, 155, 157–8, 161, 164
  rates *36*, 48, 51–2, 64–6, 79, 81, 84, 106, 110, 113, 151, 153 6
  ridge *see* spreading axes
Standard Mean Ocean Water (SMOW) 126, 161
stockwork *94*, 102, 104
strontium (Sr) 100
subduction 26–7, 31, 33, 38, 49–50, 64–5, 141, 144–6, 153, 155
  zones 27–9, 31–3, 49, 61, 63–5, 139, 144
submarine
  canyons 15, 17, 27, 30
  slide 17
  volcanoes *16*, 17, 39, 41–2, 70–2, 88–91, *88*, 157
submersibles 23–4, 39, 57, 67, 72–6, *73–4*, 78, *78*, 84, 87, 90, 98, 108
sulphate (SO$_4^{2-}$) 92, 96, 100, 102, 112, 140–1, 146, 158
sulphides (S^{2-}) 23, 92, *92*, 100, 102, 104–5, 112, 141, 158
sulphur (S) 100, 141
  dioxide (SO$_2$) 100, 139–40
Surtsey *88*
suspect terranes 62–3, 65

Tasmania 119, *119*
Tasman Sea 118–20
tests (of organisms) *117*
terrigenous sediments 57
Tertiary Period 64, 141
Tethys Ocean 53, 64–5, 134, 155
thermal
  contraction of lithosphere 35, *35*, 37, 49, 53, 88, 95
  expansion of water *56*, 121, 161
  gradient 96, 98, 113, 160

thorium (Th) 110, 160
thrust faulting 32, *33*
tidal fluctuations 121, *121*
Titanic 23
titanium (Ti) 76, 158
Tonga-Kermadec Trench 46, *47*, 152
topographical relief 43–4, *45*, *47*, 48, 81, 114, 152
trace elements 99, 107, 143
transcurrent faults 38, 58–9, 64
transform faults *25*, 36–9, *36–8*, 50, *51*, 57–8, 61, 63, 68, 74, 82–7, *82*, 90, 98, 113, 155, 157–8
transgression, marine 135–7, 163–4
Triassic Period 65
tsunamis 121
tube worms *92*
turbidity currents 30, *31*, 40, 50, 111

Udintsev Fracture Zone 48, 49, 86
unconformities 119, 129, 133
underwater photography 76, *77–8*, 90
upper mantle 48, 64, 67, *67*, 111–12, 144–5
upwelling 85
uranium (U) 110, 160

Venus, Mount 74, *74*, 76, *78*, 79, 156
vesicles 91, 157
visibility and Earth's curvature 9, *9*
volcanic
  ash 88, 90–1, 115, 157
  islands 28, 33, 41–2, *42*, 50, 86, 88–91
  see also seamounts
volcanism 31, 32–3, 39, 41–3, 45, 50, 64, 72, 78–82, *78–9*, 84–7, 90, 93, 110–11, 120, 140–1, 144, 146, 152, 156
volcanoes 28, *32*, 40–2, *41*, 49, 70–2, 75, 78, *78*, 80–2, 88–91, *88*, 153

Walvis Ridge 43
water
  budget in the Mediterranean 129–34, 143, 163
  :rock ratio 96, 100, 107–8, 112–13, 159
  warm vents 23, 92–3, *92–3*, 96, 99–105, *103*, 107–8, 110–12, 158–60
  see also hydrothermal vents
wave action 41, 121, 153
weathering *see* sea-floor weathering
Western Approaches 30
western Asia 66
western Pacific 5, 33, 48, 61, *63*, 64, 86–7, *132*
West Wind Drift *see* Antarctic Circumpolar Current
'white smokers' 92, 102–5, *103*, 112, 116, 145

zeolites 97, 108, 112–13, 159, 164
zinc (Zn) 93
  sulphide 23, 102